# Quant Probability
# 100 Interview Questions

X.Y. Wang

# Contents

1 Introduction      11

2 Basic      13

  2.1    What is probability theory and why is it important for quantitative analysts and traders? . . . . . . . . .   13

  2.2    Define the concepts of sample space, event, and outcome in the context of probability. . . . . . . . . . . .   14

  2.3    Explain the difference between discrete and continuous probability distributions. . . . . . . . . . . . . . . . .   15

  2.4    What is the law of large numbers, and why is it significant in probability theory? . . . . . . . . . . . . .   17

  2.5    Briefly describe the concept of conditional probability and provide an example. . . . . . . . . . . . . . . . .   18

  2.6    Explain the difference between dependent and independent events in probability. . . . . . . . . . . . . .   19

  2.7    Define Bayes' theorem and describe its importance in probability analysis. . . . . . . . . . . . . . . . . .   20

  2.8    What is a random variable and how is it used in probability theory? . . . . . . . . . . . . . . . . . . . . .   21

2.9 What is a probability mass function (PMF) and how is it different from a probability density function (PDF)? 22

2.10 Explain the concept of expected value and its relevance in quantitative analysis. . . . . . . . . . . . . . . . . . 24

2.11 Define variance and standard deviation, and explain their significance in measuring risk. . . . . . . . . . . . 24

2.12 Describe the normal (Gaussian) distribution and its properties. . . . . . . . . . . . . . . . . . . . . . . . . 25

2.13 Explain the central limit theorem and its importance in quantitative analysis. . . . . . . . . . . . . . . . . 27

2.14 What is the difference between covariance and correlation, and why are they important in quantitative analysis? . . . . . . . . . . . . . . . . . . . . . . . . 28

2.15 What are common probability distribution families used in quantitative finance? Provide two examples. . 29

2.16 Explain the concept of a cumulative distribution function (CDF) and its importance in probability theory. . 30

2.17 How is the Monte Carlo simulation method used in quantitative finance? . . . . . . . . . . . . . . . . . . 31

2.18 What is the significance of skewness and kurtosis in probability distributions? . . . . . . . . . . . . . . . 32

2.19 Explain the concept of the joint probability distribution and its relevance in quantitative analysis. . . . . . 33

2.20 What is a Markov chain, and how is it used in probability theory and quantitative finance? . . . . . . . . 34

3 Intermediate 37

3.1 What is the difference between a priori and empirical probabilities? Provide examples. . . . . . . . . . . . 37

3.2    How do you estimate probabilities using maximum
       likelihood estimation?  . . . . . . . . . . . . . . . . .    38

3.3    Explain the concept of marginal probability and its
       role in multivariate probability analysis.  . . . . . . . .   40

3.4    What is the law of total probability, and how is it used
       in quantitative analysis? . . . . . . . . . . . . . . . . .   41

3.5    Describe the Poisson distribution and its applications
       in quantitative finance. . . . . . . . . . . . . . . . . .    42

3.6    Explain the concept of hypothesis testing and its sig-
       nificance in quantitative analysis. . . . . . . . . . . . .   43

3.7    What is the difference between Type I and Type II
       errors in hypothesis testing? . . . . . . . . . . . . . . .   44

3.8    Describe the t-distribution and its use in hypothesis
       testing and confidence intervals. . . . . . . . . . . . . .   46

3.9    Explain the concept of a p-value and its role in hy-
       pothesis testing.  . . . . . . . . . . . . . . . . . . . .    47

3.10   Describe the Chi-squared distribution and its applica-
       tions in quantitative finance.  . . . . . . . . . . . . . .   48

3.11   Explain the concept of statistical power and its im-
       portance in hypothesis testing.  . . . . . . . . . . . . .    49

3.12   How do you apply the bootstrap method in quantita-
       tive analysis? . . . . . . . . . . . . . . . . . . . . . .    50

3.13   What are copulas and how are they used in modeling
       multivariate dependence?  . . . . . . . . . . . . . . . .     51

3.14   Explain the difference between parametric and non-
       parametric statistical methods. . . . . . . . . . . . . .     53

3.15   Describe the concept of information criteria, such as
       AIC and BIC, and their use in model selection. . . . . .      54

3.16   What is the concept of stationarity and its importance
       in time series analysis? . . . . . . . . . . . . . . . . .    55

3.17 Explain the difference between a white noise process and an autoregressive (AR) process. . . . . . . . . . . 56

3.18 Describe the concept of a moving average (MA) process and its use in time series analysis. . . . . . . . . . 57

3.19 Explain the GARCH model and its application in modeling financial time series. . . . . . . . . . . . . . 58

3.20 How do you apply principal component analysis (PCA) in quantitative finance? . . . . . . . . . . . . . . . . . 60

4 Advanced                                                        63

4.1 Explain the concept of risk-neutral probability and its relevance in option pricing. . . . . . . . . . . . . . . 63

4.2 Describe the Black-Scholes-Merton option pricing model and its key assumptions. . . . . . . . . . . . . . . . 65

4.3 How do you apply stochastic calculus in quantitative finance? Provide an example. . . . . . . . . . . . . . 66

4.4 Explain the concept of a Brownian motion and its role in stochastic processes. . . . . . . . . . . . . . . . . 67

4.5 Describe the geometric Brownian motion model and its application in option pricing. . . . . . . . . . . . 68

4.6 What is the binomial option pricing model and how does it work? . . . . . . . . . . . . . . . . . . . . . 70

4.7 Explain the concept of delta hedging and its importance in managing option risk. . . . . . . . . . . . . 71

4.8 Describe the Greeks in option pricing and their significance in risk management. . . . . . . . . . . . . . . 72

4.9 Explain the concept of implied volatility and its importance in option pricing. . . . . . . . . . . . . . . . 74

4.10 Describe the Heston model for stochastic volatility and its applications in quantitative finance. . . . . . . . . 75

4.11 What is jump diffusion modeling and its relevance in option pricing? . . . . . . . . . . . . . . . . . . . . . . 76

4.12 Explain the concept of value at risk (VaR) and its importance in risk management. . . . . . . . . . . . . 77

4.13 Describe the concept of conditional value at risk (CVaR) and how it differs from VaR. . . . . . . . . . . . . . . 78

4.14 How do you estimate portfolio risk using factor models, such as the CAPM and Fama-French models? . . . 79

4.15 What is the concept of cointegration and its application in quantitative trading strategies? . . . . . . . . . 80

4.16 Describe the concept of mean reversion in financial markets and its use in trading strategies. . . . . . . . . 81

4.17 Explain the concept of pairs trading and its application in quantitative finance. . . . . . . . . . . . . . . 82

4.18 What is the concept of algorithmic trading and its importance in modern finance? . . . . . . . . . . . . . 83

4.19 Explain the use of machine learning techniques, such as neural networks and decision trees, in quantitative finance. . . . . . . . . . . . . . . . . . . . . . . . . 84

4.20 Describe the concept of reinforcement learning and its applications in algorithmic trading strategies. . . . . . 85

5 Expert                                                                 87

5.1 Explain the concept of stochastic volatility and how it is incorporated in option pricing models. . . . . . . 87

5.2 Describe the difference between local volatility models and stochastic volatility models in option pricing. . . . 88

5.3 What is the SABR model, and how is it used in interest rate derivative pricing? . . . . . . . . . . . . . . . . . . 89

5.4 Explain the Girsanov theorem and its application in the change of measure for risk-neutral valuation. . . . 90

5.5 Describe the concept of no-arbitrage pricing and its importance in financial derivatives. . . . . . . . . . . . 92

5.6 What is the Heath-Jarrow-Morton (HJM) framework, and how is it used in interest rate modeling? . . . . . 93

5.7 Explain the concept of a forward measure and its application in fixed income securities. . . . . . . . . . . 94

5.8 Describe the Vasicek model and its application in interest rate modeling. . . . . . . . . . . . . . . . . . 95

5.9 Explain the concept of the Hull-White model and its use in interest rate derivative pricing. . . . . . . . . . 96

5.10 Describe the concept of the LIBOR Market Model (LMM) and its application in interest rate derivative pricing. . . . . . . . . . . . . . . . . . . . . . . . 97

5.11 What is the difference between risk-neutral and real-world probability measures in quantitative finance? . . 98

5.12 Explain the use of Kalman filtering in state space models for time series analysis. . . . . . . . . . . . . . 99

5.13 How do you apply advanced optimization techniques, such as genetic algorithms and simulated annealing, in portfolio optimization? . . . . . . . . . . . . . . . . 100

5.14 Describe the concept of market microstructure and its relevance in high-frequency trading. . . . . . . . . . 101

5.15 Explain the role of limit order books and order flow in market microstructure research. . . . . . . . . . . . 102

5.16 What is the concept of optimal execution in algorithmic trading, and what are some common strategies? . 103

5.17 Describe the use of text analysis and natural language processing (NLP) in quantitative finance. . . . . . . . 104

5.18   Explain the concept of regime-switching models and
       their application in financial markets. . . . . . . . . . 105

5.19   How do you incorporate transaction costs and other
       frictions in portfolio optimization and trading strategies? 107

5.20   What are some common methods for evaluating the
       performance of quantitative trading strategies, and how
       do you account for overfitting? . . . . . . . . . . . . . 108

6  Guru                                                          111

6.1    Describe the concept of rough volatility models and
       their application in option pricing. . . . . . . . . . . . 111

6.2    Explain the use of model-free implied volatility mea-
       sures, such as the VIX index, in quantitative finance. . 112

6.3    How do you incorporate market frictions, such as liq-
       uidity and transaction costs, in derivative pricing models? 113

6.4    Describe the concept of a change of numéraire and its
       application in fixed income and derivative pricing. . . 114

6.5    Explain the martingale representation theorem and its
       significance in quantitative finance. . . . . . . . . . . 116

6.6    What is the concept of affine term structure models,
       and how are they used in interest rate modeling? . . . 117

6.7    Describe the use of neural network-based calibration
       methods for complex derivative pricing models. . . . . 118

6.8    Explain the concept of utility maximization and its
       role in optimal portfolio selection. . . . . . . . . . . . 119

6.9    How do you incorporate higher moments, such as
       skewness and kurtosis, in portfolio optimization and
       risk management? . . . . . . . . . . . . . . . . . . . . 120

6.10   Describe the concept of stochastic control and its ap-
       plication in algorithmic trading and risk management. 122

6.11  What is the role of information theory in quantitative finance, and how is it applied to model selection and trading strategies? . . . . . . . . . . . . . . . . . . .  123

6.12  Explain the concept of dynamic copula models and their application in modeling multivariate dependence in financial markets. . . . . . . . . . . . . . . . . . . .  124

6.13  How do you apply the concept of market incompleteness in derivative pricing and risk management? . . . .  125

6.14  Describe the use of fractional Brownian motion and long memory processes in financial modeling. . . . . .  126

6.15  What is the role of behavioral finance in quantitative trading strategies, and how can it be incorporated into models? . . . . . . . . . . . . . . . . . . . . . . . . .  127

6.16  Explain the concept of ambiguity aversion and its implications for financial decision-making and modeling.  128

6.17  How do you incorporate systemic risk factors in portfolio optimization and risk management? . . . . . . . .  130

6.18  Describe the concept of optimal stopping and its applications in financial derivatives and trading strategies. 131

6.19  Explain the use of agent-based models in simulating complex market dynamics and their applications in quantitative finance. . . . . . . . . . . . . . . . . .  133

6.20  What are the challenges and limitations of applying machine learning and artificial intelligence techniques in quantitative finance, and how can they be mitigated? 134

# Chapter 1

# Introduction

The field of quantitative finance has become increasingly important in the last few decades as the complexity of financial markets and instruments has grown. Alongside this development, the application of probability theory has emerged as a crucial component for quantitatively analyzing and understanding the financial markets. The purpose of this book, "Quant Probability 100 Interview Questions", is to provide a comprehensive resource for those who are looking to deepen their knowledge of probability theory and its applications in quantitative finance. Whether you are a student, a professional, or an enthusiast of quantitative finance, this book will help to demystify probability theory and equip you with the knowledge to excel in this field.

The book is organized into five sections, ranging from basic to guru level, and covers a wide array of topics in probability and quantitative finance. Starting with foundational concepts in probability theory, the book advances through intermediate and advanced topics, culminating in expert and cutting-edge applications of probability in quantitative finance. Each section consists of carefully selected interview questions that highlight essential aspects of probability theory and its applications. These questions are accompanied by detailed explanations, examples, and further resources to facilitate a thorough understanding of the concepts.

In the Basic section, you will learn about fundamental probability concepts, such as sample space, events, and random variables. You will also be introduced to the key principles of probability theory, such as the law of large numbers, Bayes' theorem, and conditional probability. Additionally, this section delves into various probability distributions, statistical measures, and the use of Monte Carlo simulation in quantitative finance.

The Intermediate section builds upon the foundation established in the Basic section, covering topics such as hypothesis testing, statistical power, copulas, time series analysis, and various probability distribution families commonly used in quantitative finance. You will also learn about maximum likelihood estimation and the bootstrap method, among other essential techniques in quantitative analysis.

In the Advanced section, the focus shifts to more specialized topics and applications of probability theory in quantitative finance. This includes options pricing models, stochastic calculus, GARCH models, risk management, algorithmic trading, and the use of machine learning techniques in finance.

The Expert section delves into the intricacies of derivative pricing, interest rate modeling, stochastic volatility, market microstructure, and advanced optimization techniques. You will also explore topics such as the Girsanov theorem, no-arbitrage pricing, and the use of natural language processing in quantitative finance.

Lastly, the Guru section covers cutting-edge concepts and techniques in probability theory and quantitative finance, including rough volatility models, utility maximization, stochastic control, behavioral finance, and agent-based models. This section also addresses the challenges and limitations of applying machine learning and artificial intelligence techniques in quantitative finance and suggests ways to mitigate them.

"Quant Probability 100 Interview Questions" is a comprehensive guide for anyone seeking to advance their understanding of probability theory and its applications in quantitative finance. Through a carefully curated selection of interview questions and in-depth explanations, this book will help you master the essential concepts and techniques required to excel in this exciting and dynamic field.

# Chapter 2

# Basic

## 2.1 What is probability theory and why is it important for quantitative analysts and traders?

Probability theory is a branch of mathematics that deals with the analysis of random phenomena. It provides a framework for quantifying and evaluating the likelihood of events based on the knowledge of their underlying factors. Probability theory forms a fundamental basis for statistical inference, decision-making, and risk management.

Probability theory is essential for quantitative analysts and traders because it enables them to understand the uncertain nature of financial markets and make informed trading decisions. By modeling the probabilities of different market outcomes, analysts can develop trading strategies that optimize risk and return. This allows them to determine the optimal allocation of capital among investments, manage their portfolios, and respond to market changes.

For example, a quantitative analyst might use probability theory to calculate the likelihood that a particular stock will change in value, and use that information to design a trading strategy. They might analyze market trends, identify risk factors, and model various scenarios to understand the potential impact of market conditions on their investment portfolios.

Probability theory is also important for risk management in trading. By calculating probabilities of various market outcomes, traders can assess risk levels and develop strategies to minimize potential losses. For instance, they might use stop loss orders to limit losses on a particular position or diversify their holdings to reduce exposure to specific risks.

In order to implement quantitative trading strategies, quantitative analysts and traders use various mathematical and statistical tools, such as stochastic calculus, Monte Carlo simulations, and time series analysis. They use programming languages such as Java, R, and Python to implement these methods, and are able to test and refine their models using historical data.

## 2.2   Define the concepts of sample space, event, and outcome in the context of probability.

In probability theory, a sample space is the set of all possible outcomes of a random experiment. For example, if we roll a six-sided die, the sample space is 1, 2, 3, 4, 5, 6. This means that any time we roll the die, the outcome will be one of these six values.

An event is a subset of the sample space that we are interested in studying. For example, if we are only interested in the outcomes that are greater than 3, the event we are studying is 4, 5, 6. We can also study more complex events that require multiple conditions to be met, such as rolling two dice and looking for the sum to be greater than 9.

An outcome, also called a sample point, is a particular result of the random experiment. For example, if we roll the die and it comes up showing 3, the outcome is 3. It is important to note that an outcome must be a member of the sample space, otherwise it is not a valid outcome.

In summary, sample space is the set of all possible outcomes, event is a subset of the sample space that we are interested in studying, and an outcome is a particular result of the random experiment. These three concepts are the foundation of probability theory and are used to define mathematical models of uncertainty and make quantitative predictions in trading and investment.

## 2.3 Explain the difference between discrete and continuous probability distributions.

Probability distributions are mathematical representations of the probability of different outcomes in an experiment or event.

A discrete probability distribution is a probability distribution that deals with a countable number of outcomes. In other words, it is a distribution that deals with variables that can only take on certain values. These variables can only take on values that are distinct and separate from each other. For example, the outcome of rolling a die is discrete because there are only six possible outcomes: 1, 2, 3, 4, 5 or 6.

On the other hand, a continuous probability distribution is a probability distribution that deals with variables that can take on any value within a given interval. This means that the outcomes are not discrete and can include all possible values within the interval. Continuous distributions are often associated with measurements and physical phenomena. For example, the height of a person is a continuous distribution because it can take on any value within a given range.

In terms of quantitative trading/investment, probability theory is applied to model the behavior of financial instruments such as stocks, options, and futures. Discrete probability distributions are often used to model the behavior of discrete financial events such as stock prices that only change in small increments. For example, a binomial distribution can be used to model the price change of a stock that can either go up or down by a certain amount.

Continuous probability distributions, on the other hand, are often used to model the behavior of continuous financial events such as interest rates or returns. For example, a normal distribution can be used to model the returns of a portfolio of stocks over time.

## 2.4    What is the law of large numbers, and why is it significant in probability theory?

The Law of Large Numbers (LLN) is a principle in probability theory that describes the behavior of the average of a large number of independent and identically distributed (i.i.d.) random variables. This law states that as the number of samples (or observations) in a random variable increases, the average of the observed values approaches the expected value of the variable.

In other words, if we randomly sample a large number of observations from the same probability distribution, then the average of those observations will eventually converge to the expected value of that same distribution. Thus, the larger the sample size, the more accurate the average value will be in reflecting the true nature of the underlying probability distribution.

The significance of LLN in probability theory is that it provides a foundation for many fundamental statistical techniques, such as hypothesis testing, confidence intervals, and regression analysis. For example, the Central Limit Theorem, which is a key theorem in probability theory and statistics, relies on the LLN to hold. The theorem states that for any distribution with a finite variance, the sum of a large number of random variables (with finite mean and variance) approaches a normal distribution as the sample size increases.

In quantitative trading/investment, the LLN has significant implications for managing investment risk. According to the LLN, as the number of observations increases, the variability around the expected value of the investment returns decreases. For example, if we are analyzing the historical returns of a stock market index over a long period, then we can expect that the average annual returns will converge to the average expected returns of that index over time, reducing the risk of investing in the index.

In summary, the LLN is significant in probability theory and quantitative trading/investment because it provides a foundation for many statistical techniques and reduces the risk of investment by reducing the variability of returns around the expected value of an investment.

## 2.5 Briefly describe the concept of conditional probability and provide an example.

Conditional probability is a statistical measure that calculates the likelihood of an event occurring based on the occurrence of a previous event. Specifically, it is the probability of a particular outcome given that another event has already occurred.

It can be expressed mathematically as:

$$P(A|B) = P(A \wedge B)/P(B) \tag{2.1}$$

Where P(A) represents the probability of event A occurring, P(B) is the probability of event B occurring, $P(A|B)$ is the probabiity of event A occuring given evnet B occured, and $P(A \wedge B)$ is the probability of both A and B occurring.

For example, consider a bag that contains three red marbles and two blue marbles. If we randomly draw one marble from the bag, the probability of selecting a red marble is 3/5 or 0.6. However, if we draw a marble from the bag and know that it is blue, the probability of the next marble being red becomes conditional on the first result.

If we drew a blue marble in our first draw and then replaced it, the probability of drawing a red marble on the second draw would be:

$$P(Red|Blue) = P(Red \wedge Blue)/P(Blue) \tag{2.2}$$

Since the marble was replaced after the first draw, P(Blue) = 2/5.

The probability of drawing a red and blue marble is 3/5 * 2/5 = 6/25. Therefore:

$$P(Red|Blue) = (3/5 * 2/5)/(2/5) = 3/5 \qquad (2.3)$$

This means that the probability of drawing a red marble on the second draw, given that we drew a blue marble on the first draw, is 0.6.

## 2.6   Explain the difference between dependent and independent events in probability.

In probability theory, events are classified as either dependent or independent. These terms are used to describe the relationship between two or more events and their outcomes.

Independent events are those where the occurrence of one event has no effect on the occurrence of the other. The probability of one event occurring does not affect the probability of the other event occurring. For example, when flipping a coin, getting heads on the first flip does not affect the probability of getting heads on the second flip. Each flip is independent of the other.

Dependent events, on the other hand, are those events where the occurrence of one event does affect the probability of the other event occurring. The outcome of one event is dependent on the outcome of the other. For example, drawing two cards from a deck without replacement is a dependent event. The probability of drawing a second card of a certain suit after drawing a card of that suit on the first draw is affected by the removal of that card from the deck.

To illustrate this difference, let's consider the example of rolling two dice:

- Independent events: Rolling two dice and getting a 4 on the first die and a 3 on the second die are independent events. The probability of rolling a 4 on the first die is 1/6, and the probability of rolling a 3 on the second die is also 1/6. The probability of rolling a 4 on the first die does not affect the probability of rolling a 3 on the second die, so the probability of rolling a 4 on the first die and a 3 on the second die is the product of their individual probabilities, which is 1/36.

- Dependent events: Rolling two dice and getting a sum of 7 is a dependent event. The outcome of the first roll affects the probability of the second roll. If we roll a 1 on the first die, then the probability of rolling a 6 on the second die to get a sum of 7 is 1/6. However, if we roll a 5 on the first die, then the probability of rolling a 2 on the second die to get a sum of 7 is 0, as it is not possible. Therefore, the probability of rolling a sum of 7 is not simply the product of the probabilities of rolling a 1 and a 6 or rolling a 2 and a 5, but rather depends on the possible combinations of rolls that result in a sum of 7.

## 2.7    Define Bayes' theorem and describe its importance in probability analysis.

Bayes' theorem is a fundamental concept in probability theory that allows us to update or revise our estimate of the probability of an event occurring based on new information or evidence. It is named after Reverend Thomas Bayes, an 18th-century mathematician who first introduced the theorem.

The formula for Bayes' theorem is as follows:

$P(A|B) = (P(B|A) * P(A)) / P(B)$

Where:

  $P(A)$ is the prior probability of event A

  $P(B)$ is the prior probability of event B

  $P(A|B)$ is the posterior probability of event A given B

  $P(B|A)$ is the conditional probability of event B given A

In other words, Bayes' theorem allows us to update our prior belief about the probability of event A occurring, based on the information or evidence provided by event B. We use the conditional probability $P(B|A)$ to determine the likelihood of observing B given that A has occurred, and then multiply it by the prior probability P(A) to obtain the joint probability of A and B.

Finally, we divide this joint probability by the prior probability of B (P(B)) to normalize it and obtain the posterior probability of A given B, which is the updated probability estimate taking into account the new evidence.

Bayes' theorem is important in probability analysis because it provides a rigorous and systematic approach to updating our probability estimates based on new information, and can be used to model a wide range of real-world scenarios such as medical diagnoses, financial investments, and climate predictions. It also allows us to quantitatively evaluate the strength of evidence in support of a hypothesis or theory, and to make more informed decisions based on probabilistic reasoning.

In quantitative trading and investment, Bayes' theorem can be used to update our beliefs about the future prospects of a company or asset based on new information such as earnings reports or economic data. For example, if we initially believe that a company is likely to be profitable based on its past performance and industry trends, but then receive negative news about its financial health, we can use Bayes' theorem to revise our probability estimate of its profitability in light of this new information. This can help us make more informed investment decisions and minimize our exposure to risk.

## 2.8   What is a random variable and how is it used in probability theory?

In probability theory, a random variable is a quantity whose value depends on the outcome of a random event or experiment. It is a mathematical function that maps the possible outcomes of an experiment to numerical values. Essentially, a random variable is a way of quantifying uncertainty and converting it into mathematical language.

Random variables can be classified into two types: discrete and continuous. A discrete random variable takes on a finite or countably infinite set of values, such as coin tosses or dice rolls. A continuous random variable, on the other hand, takes on an uncountably infinite set of values within a specified range, such as the height of a randomly selected person or the amount of rainfall in a specific region.

Random variables are used in probability theory to define probability distributions, which describe the likelihood of different outcomes of an experiment or event. Probability distributions for discrete random variables are typically expressed as probability mass functions (PMFs), while probability distributions for a continuous random variable are typically expressed as probability density functions (PDFs).

The use of random variables is crucial in many areas of quantitative trading and investment. They are commonly used in quantitative finance to model the behavior of financial markets and instruments, such as stock prices, interest rates, and currency exchange rates. Random variables also play a central role in risk management, where they are used to model the likelihood of extreme events and potential losses that can occur in financial markets. By using probabilistic models based on random variables, traders and investors can make informed decisions about when and how to trade or invest, based on the expected returns, risk and uncertainty associated with different financial instruments.

## 2.9   What is a probability mass function (PMF) and how is it different from a probability density function (PDF)?

A probability mass function (PMF) and a probability density function (PDF) are both ways of describing the probability distribution of a random variable. However, they are different in how they are defined and the types of random variables they describe.

A probability mass function (PMF) is a function that gives the probability that a discrete random variable takes on a certain value. That is, for each possible value of the random variable, the PMF gives the probability of that value occurring. The PMF is typically denoted by $P(X = x)$, where X is the random variable and x is a possible value it can take on. The PMF satisfies two important properties:

1. Non-negativity: $P(X = x) \leq 0$ for all x.

2. Normalization: $\sum P(X = x) = 1$ over all possible values of X.

For example, let X be the result of rolling a fair six-sided die. The PMF of X is $P(X = 1) = P(X = 2) = P(X = 3) = P(X = 4) = P(X = 5) = P(X = 6) = 1/6$.

On the other hand, a probability density function (PDF) is a function that gives the relative likelihood of observing a continuous random variable within a certain range of values. That is, for a continuous random variable, the PDF gives the probability density at various points along the continuous range of possible values, but not the probability of any specific value occurring. The PDF is typically denoted by f(x). The properties that a PDF satisfies are:

1. Non-negativity: $f(x) > 0$ for all x.

2. Normalization: $\int f(x)dx = 1$ over the entire range of x.

For example, let X be a continuous random variable that represents the length of a certain type of fish in a lake, measured in inches. A possible PDF for X is $f(x) = 0.5 * e^{(-0.5*x)}$ for $x > 0$, where e represents the mathematical constant e 2.718. This PDF satisfies the two properties listed above and gives the relative likelihood of observing a fish of a given length.

In summary, PMFs are used to describe the probability of observing specific values for a discrete random variable, while PDFs are used to describe the relative likelihood of observing continuous values for a continuous random variable.

## 2.10  Explain the concept of expected value and its relevance in quantitative analysis.

Expected value is a fundamental concept in probability theory that measures the long-term average value of a random variable. It is calculated by multiplying each possible outcome by its probability and summing these values.

In finance, expected value is used to determine the expected return on an investment, which is the potential gain or loss on an investment weighted by its probability of occurring. This is a critical concept in quantitative analysis, as it allows traders and investors to make decisions based on the expected value of different investment opportunities.

For example, suppose a stock has a 60% chance of increasing in value by 10% and a 40% chance of decreasing in value by 5%. The expected value of this investment would be calculated as follows:

Expected value = (0.6 x 0.10) + (0.4 x -0.05) = 0.06 - 0.02 = 0.04

This means that, on average, the investment is expected to provide a return of 4%. Knowing this information can help an investor decide whether or not to invest in this stock compared to other investment opportunities, as they can evaluate the expected value of each option.

Overall, expected value is an essential part of quantitative analysis in finance, as it is used to evaluate the potential returns and risks of different investment opportunities.

## 2.11 Define variance and standard deviation, and explain their significance in measuring risk.

Variance and standard deviation are two important statistical measures used to quantify the risk associated with investment returns.

Variance is often used to describe the degree of spread or dispersion of a set of data points. In the context of investment returns, variance provides a measure of how much the actual returns deviate from the expected or average returns. Specifically, variance is calculated by finding the average of the squared differences between each data point and the mean. The formula for variance is as follows:

$$Variance = \sqrt{(\sum(x_i - x)^2)/(n - 1))} \tag{2.4}$$

where $x_i$ is each data point, x is the mean, and n is the total number of data points.

Standard deviation, on the other hand, is the square root of variance. It measures the amount of dispersion of a set of data from its mean, expressed in the same units as the data. Standard deviation is more commonly used than variance as it provides a more easily understandable measure of risk.

In investment terms, standard deviation provides a measure of the volatility or uncertainty of investment returns. A higher standard deviation indicates a wider range of possible outcomes, which means there is a greater potential for both positive and negative returns. Conversely, a lower standard deviation indicates a narrower range of possible outcomes, which means there is less uncertainty and generally less risk involved.

In summary, variance and standard deviation provide a quantitative way to measure the level of risk associated with investment returns. Knowing these statistical measures can help investors understand the potential risks and rewards of different investment options, and make more informed decisions.

## 2.12    Describe the normal (Gaussian) distribution and its properties.

The normal distribution, also known as the Gaussian distribution, is a probability distribution that describes the behavior of a random variable that is continuously distributed. It is one of the most important probability distributions in the fields of finance, statistics, and science. The normal distribution is characterized by its bell-shaped curve, which is symmetric around its mean.

The properties of the normal distribution are:

1. Mean: The mean of a normal distribution represents the central

tendency of the data. It is equal to the highest point on the curve, also known as the peak or mode. The mean of a normal distribution is also the location of the horizontal line of symmetry that divides the distribution in half.

2. Standard Deviation: The standard deviation of a normal distribution represents the spread of the data. It is a measure of how much the data deviates from the mean. The standard deviation controls the width of the bell-shaped curve. The larger the standard deviation, the wider the curve will be.

3. Skewness: A normal distribution has zero skewness. This means that the curve is symmetric around its mean, and the tails of the curve extend equally to the left and right.

4. Kurtosis: A normal distribution has a kurtosis of three. This means that the curve is neither flat nor peaked compared to the bell curve. It has a moderate level of peakedness.

5. Empirical Rule: The normal distribution follows the empirical rule, also known as the 68-95-99.7 rule. This rule states that approximately 68% of the data falls within one standard deviation of the mean, 95% falls within two standard deviations, and 99.7% falls within three standard deviations.

The normal distribution is widely used in quantitative trading and investment management as it provides a powerful mathematical tool for analyzing and modeling market data. It can be used to calculate probabilities of future market events, to estimate trading risk and return, and to generate trading signals.

## 2.13   Explain the central limit theorem and its importance in quantitative analysis.

The central limit theorem is a fundamental concept in probability theory and statistics that describes the behavior of the averages of a large number of independent and identically distributed random variables. More specifically, it states that the sum of a large number of independent and identically distributed random variables roughly follows a normal distribution, regardless of the distribution of the individual variables.

In simple terms, the central limit theorem tells us that as we take larger and larger sample sizes, the sample means tend to converge to a normal distribution, even if the original population from which the samples are drawn is not normally distributed. This theorem is extremely important in quantitative analysis because it allows us to make accurate inferences about the population based on a limited amount of data.

For example, suppose we are interested in estimating the average height of all people in a particular country. We could collect a large number of height measurements and calculate the average, but this is often impractical. Instead, we could take a random sample of people from the population and measure their height. According to the central limit theorem, as our sample size increases, our estimate of the population mean becomes more and more accurate, even if the distribution of heights within the population is not normal.

The central limit theorem is also important in quantitative trading and investment. Many financial models and trading strategies rely on the assumption that the returns of different assets follow a normal distribution. The central limit theorem provides a justification for this assumption, as it suggests that the sum of many small, independent market forces can produce normally distributed outcomes.

Overall, the central limit theorem is a critical tool for probabilistic analysis in a wide range of fields, including finance, economics, engineering, and scientific research. It allows us to make accurate predictions and inferences about large, complex systems based on limited data, which is essential for informed decision-making.

## 2.14   What is the difference between co-variance and correlation, and why are they important in quantitative analysis?

Covariance and correlation are two important measures of the relationship between two variables in quantitative analysis.

Covariance is a measure of how two variables vary together. It is calculated as the average of the product of the deviations of each variable from its mean. A positive covariance indicates that the variables tend to move in the same direction, while a negative covariance indicates that they tend to move in opposite directions.

Correlation, on the other hand, measures the strength and direction of the linear relationship between two variables. It is a standardised measure that ranges from -1 to +1. A correlation of +1 indicates a perfect positive relationship, a correlation of -1 indicates a perfect negative relationship, and a correlation of 0 indicates no linear relationship.

The main difference between covariance and correlation is that covariance is not standardised and can take on any value, while correlation is standardised and always falls between -1 and +1. This makes correlation a more useful measure for comparing the strength of relationships between variables with different scales.

Both covariance and correlation are important in quantitative analysis because they provide information about the degree and direction of the relationship between two variables. This information can be used to help identify patterns and make predictions about future behaviour. For example, in finance, covariance and correlation are used to measure the relationship between different assets and to construct portfolios that minimise risk while maximising returns. In medical research, correlation is used to measure the strength of association between risk factors and specific diseases.

In summary, while covariance and correlation are both measures of the relationship between two variables, correlation is a more useful measure for comparing the strength of relationships between variables with different scales. Both measures are important in quantitative analysis for identifying patterns and making predictions.

## 2.15 What are common probability distribution families used in quantitative finance? Provide two examples.

There are several probability distribution families that are commonly used in quantitative finance, depending on the specific problem and the underlying assumptions. Two examples of probability distribution families that are frequently used in quantitative finance are the normal (or Gaussian) distribution and the Student's t-distribution.

The normal distribution is often used in finance to model the behavior of stock prices, interest rates, and other financial variables. The normal distribution is characterized by its bell-shaped curve and its two parameters: mean (denoted ) and standard deviation (denoted ). The mean represents the center of the distribution, while the standard deviation represents its width. The probability density function (PDF) of the normal distribution is given by:

$$f(x) = (1/\sigma\sqrt{2\pi}) * \exp(-(x-\mu)^2/(2\sigma^2)), \qquad (2.5)$$

where x represents the random variable, $\mu$ is the mean, $\sigma$ is the standard deviation, and $\pi$ the mathematical constant.

An example where the normal distribution is used in quantitative finance is in the Black-Scholes-Merton model, which is a widely used mathematical model for pricing options. In this model, the underlying asset's price is assumed to follow a log-normal distribution, which is a variant of the normal distribution.

The Student's t-distribution is another probability distribution family that is commonly used in quantitative finance. The t-distribution is often used when the sample size is small or when the population standard deviation is unknown. The t-distribution has a similar shape to the normal distribution, but with heavier tails, which means that it is more likely to observe extreme values or outliers. The t-distribution has three parameters: mean (denoted ), scale (denoted ), and degrees of freedom (denoted ), which determines the shape of the distribution. The PDF of the t-distribution is given by:

$$f(x) = \frac{\Gamma((n+1)/2}{\Gamma(n/2)} \frac{1}{\sqrt{n\pi}} \left(1 + (x-\mu)^2/n\right)^{(-(n+1)/2)}, \qquad (2.6)$$

where x represents the random variable, $\mu$ is the mean, n is the degrees of freedom, $\tau$ is the gamma function, and $\pi$ is the mathematical constant.

An example where the t-distribution is used in quantitative finance is in the estimation of VaR (Value at Risk), which is a statistical measure of the maximum potential loss that an investment portfolio may suffer within a given time frame, with a certain level of confidence. The t-distribution is used to generate the critical values of VaR under the assumption that the returns of the assets in the portfolio are normally distributed.

## 2.16 Explain the concept of a cumulative distribution function (CDF) and its importance in probability theory.

The cumulative distribution function (CDF) is a fundamental concept in probability theory. It describes the probability that a random variable takes a value less than or equal to a particular number. In other words, the CDF of a random variable X is defined as F(x) = P(X<x), where x is any real number.

The CDF provides a complete characterization of the distribution of a random variable. It is particularly important for continuous random variables because it enables us to calculate the probability that a random variable takes on a value within a certain interval. For example, suppose we have a continuous random variable X that represents the heights of adult males in a certain population. The CDF F(x) for X specifies the probability that a randomly selected male is less than or equal to x inches tall. Using this information, we can calculate the probability that a randomly selected male is between a certain range of heights, such as between 5'8" (68 inches) and 6'2" (74 inches).

The CDF is also useful for identifying percentiles of a distribution. For example, the 90th percentile of X represents the point at which 90% of the population has heights less than or equal to that value.

In quantitative trading and investment, the CDF is commonly used to model the probability of large price movements in financial markets. The CDF of daily returns can help traders and investors understand the risk and potential payoff of different investment strategies. For example, a trader may use the CDF to calculate the probability of experiencing a large loss or gain on a particular day, and adjust their trading strategy accordingly.

## 2.17    How is the Monte Carlo simulation method used in quantitative finance?

Monte Carlo simulation is a popular statistical method used in quantitative finance to model financial assets and portfolio behaviors. It is widely used to estimate the likelihood of particular financial events, the probabilities of various portfolio outcomes, and the overall risk of individual financial instruments or portfolios.

In the context of finance, Monte Carlo simulation involves building a computational model of a financial asset or portfolio, which includes parameters that can take on a range of values. The analyst then randomly generates a large number of simulations based on different scenarios, using these parameters to estimate the probability of different financial outcomes.

For instance, Monte Carlo simulation can be used to model the potential risks and returns of an investment portfolio over a particular period. In order to do this, the analyst will specify inputs such as the portfolio's expected returns, standard deviation, and correlation with other asset classes. They can also simulate external events such as market crashes or geopolitical shocks that could influence the portfolio's value.

The Monte Carlo simulation then generates a set of possible returns for the portfolio based on the specified inputs and generates a probability distribution of returns. This distribution provides insight into the likelihood of different possible outcomes for the portfolio's value, allowing the investor to assess their risk relative to their desired outcome. Through this analysis, investors can identify potential opportunities to optimize their portfolio and manage their risk exposure.

Overall, Monte Carlo simulation is highly beneficial within quantitative finance because it allows analysts and investors to simulate a wide range of financial events and outcomes, estimate the probabilities of each scenario, and optimize financial strategies accordingly. It is a powerful tool that can help investors take a more informed and strategic approach to investment decisions.

## 2.18   What is the significance of skewness and kurtosis in probability distributions?

Skewness and kurtosis are important characteristics of probability distributions that provide information about the shape and behavior of the distribution.

Skewness measures the degree of symmetry of a distribution. A perfectly symmetrical distribution has a skewness of 0. A positive skewness indicates that the tail of the distribution is longer on the positive side than on the negative side, while a negative skewness indicates the opposite. In finance and investing, skewness is important because it can indicate the potential for extreme outcomes. For example, a positively skewed distribution may indicate that there is a greater likelihood of experiencing large losses than large gains.

Kurtosis, on the other hand, measures the degree of peakedness or flatness of a distribution. A normal distribution, with a bell-shaped curve, has a kurtosis of 3. A distribution with a kurtosis greater than 3 is said to be leptokurtic, indicating that it has more scores in the tails than a normal distribution. A distribution with a kurtosis less than 3 is said to be platykurtic, indicating that it has fewer scores in the tails than a normal distribution. High kurtosis distributions are

important in finance and investing because they can indicate periods
of high volatility and unexpected events that can have significant
impact on the underlying asset.

In summary, skewness and kurtosis are important characteristics of
probability distributions that provide valuable information about the
shape and behavior of the distribution. Finance and investing pro-
fessionals can use these measures to better understand the potential
risks and outcomes associated with different investment strategies and
asset classes.

## 2.19  Explain the concept of the joint prob-
## ability distribution and its relevance
## in quantitative analysis.

In probability theory and statistics, a joint probability distribution
is a probability distribution that describes the likelihood of two or
more random variables taking specific values simultaneously. It is
a function that assigns each possible combination of values of the
random variables with a probability value.

For example, let's suppose that we have two random variables, A
and B, both of which can take on the values of 1, 2 or 3. The joint
probability distribution corresponds to the probability of A taking
on a particular value and B taking on another particular value. The
following table shows the possible combinations of A and B, along
with their corresponding probabilities:

```
|       | B = 1  | B = 2  | B = 3  |
|---|-------|-------|-------|
| A = 1 | P(A=1, B=1) | P(A=1, B=2) | P(A=1, B=3) |
| A = 2 | P(A=2, B=1) | P(A=2, B=2) | P(A=2, B=3) |
| A = 3 | P(A=3, B=1) | P(A=3, B=2) | P(A=3, B=3) |
```

The joint probability distribution is important in quantitative analysis because it allows us to study the relationship between two or more
random variables. By analyzing the joint probabilities, we can determine if two variables are independent, meaning that changes in one
variable do not affect the likelihood of the other variable taking on

a certain value. Or we can determine if the variables are dependent, such that changes in one variable are associated with changes in the likelihood of the other variable taking on a certain value.

Furthermore, we can use the joint probability distribution to calculate the marginal distributions of individual variables. The marginal distribution of a single variable is the probability distribution of that variable by itself, regardless of the value of any other variables.

Finally, the joint probability distribution can be used to calculate various statistical measures such as covariance, correlation, and conditional probability. These measures help us to understand the relationship between variables and provide valuable insights for making investment decisions or creating quantitative trading strategies.

## 2.20 What is a Markov chain, and how is it used in probability theory and quantitative finance?

A Markov chain is a mathematical model used to describe the probabilistic transitions of a random process that changes over time. It is a type of stochastic process in which the likelihood of a future event is based only on the current state of the system and not on any previous history.

In a Markov chain, the system being modeled consists of a set of states and a probability of transitioning from one state to another. The probabilities of transitioning between states are represented by a transition matrix, which defines the likelihood of moving from one state to another in a given period of time.

Markov chains are commonly used in probability theory to model a wide variety of processes, including the behavior of financial markets. In quantitative finance, Markov chain models are used to forecast asset returns and identify trends and patterns in markets. These models can also be used to create trading strategies that take advantage of market inefficiencies and identify profitable opportunities.

One of the most significant advantages of using a Markov chain in financial modeling is its ability to incorporate information about the current state of the market into predictions about the future. By using a transition matrix that incorporates current market conditions, financial analysts can create a more accurate and reliable model of market behavior.

For example, a trader might use a Markov chain model to identify profitable trades based on patterns in the behavior of a particular asset. By analyzing the transition matrix for that asset, they could identify trends and patterns that suggest the price of the asset is likely to increase or decrease in the near future. This information could then be used to execute a profitable trade.

In summary, Markov chains are an essential tool in both probability theory and quantitative finance. They allow analysts to model complex systems and predict the behavior of financial markets with greater accuracy and reliability.

# Chapter 3

# Intermediate

## 3.1 What is the difference between a priori and empirical probabilities? Provide examples.

A priori probabilities and empirical probabilities are two different types of probabilities used in probability theory and quantitative trading/investment.

A priori probability is a probability that is determined or calculated based on knowledge, reasoning or logic, rather than being based on observation or data. A priori probabilities are often used in theoretical probability, and represent the likelihood or chance of an event occurring based on logical or deductive reasoning.

For example, if we toss a fair six-sided die, the probability of rolling a 5 is 1/6. This is an a priori probability because we can logically determine the probability based on the knowledge that there are six possible outcomes, each equally likely to occur.

Empirical probability, on the other hand, is a probability that is based on observation or data. Empirical probabilities are often used in experimental or empirical studies, such as in finance, economics, and other sciences that require the use of statistical analysis.

For example, if we conduct an experiment by tossing a coin 100 times, and the coin lands on heads 60 times, we can say that the empirical probability of getting heads is 0.6 or 60%. This is an empirical probability because we are calculating the probability based on actual observed data.

In summary, a priori probabilities are calculated based on logical and theoretical reasoning, while empirical probabilities are calculated based on actual observed data. Both types of probabilities are useful in probability theory and in quantitative trading/investment strategies, as they can provide insight into the likelihood of events occurring and help to inform decision-making processes.

## 3.2    How do you estimate probabilities using maximum likelihood estimation?

Maximum likelihood estimation (MLE) is a statistical technique used to estimate the parameters of a probability distribution based on a set of observed data. It is widely used in quantitative trading and investment, particularly in the estimation of expected returns and risk metrics.

The principle of MLE is to find the parameter values that maximize the likelihood of observing the data. The likelihood function is the probability of observing the data given a particular set of parameter values. The optimal parameter values are those that result in the highest likelihood function value.

To estimate probabilities using MLE, we need to follow the following steps:

1. Define the statistical model: We first need to define the probability distribution that we want to estimate the parameters for. This involves selecting an appropriate distribution that best describes the data, such as Normal, Poisson, or Binomial distribution.

2. Write down the likelihood function: The likelihood function is the joint probability density function of the data given the parameters.

For example, if we assume that the data is Normally distributed, the likelihood function can be written as:

$$L(\mu, \sigma | x_1, x_2, ..., x_n) = (2\pi\sigma^2)^{-n/2} exp\left[-\sum(1/2\sigma^2)(x_i - \mu)^2\right]$$
$$(3.1)$$

where $\mu, \sigma$ represents the parameters of the normal distribution (mean and square root of variance), x is the observed data, and n is the sample size.

3. Take the derivative of the likelihood function: We can use calculus to find the maximum likelihood estimates of the parameters by taking the derivative of the likelihood function with respect to the parameter(s) and setting it equal to zero. This will give us the values of the parameters that maximize the likelihood of observing the data.

4. Solve for the parameter(s): Solving for the parameter(s) involves finding the roots of the derivative of the likelihood function. In some cases, the closed-form solution can be found analytically, while in other cases, numerical optimization techniques such as gradient descent or Newton-Raphson method are used.

5. Evaluate the estimated probabilities: Once we have estimated the parameters of the probability distribution, we can use them to estimate the probabilities of future events or outcomes. For example, if we have estimated the parameters of a Normal distribution, we can use them to calculate the probability of observing a certain value or range of values in the future.

In summary, maximum likelihood estimation is a powerful statistical technique that can be used to estimate the parameters of probability distributions, which can then be used to estimate the probabilities of future events or outcomes. It is widely used in quantitative trading and investment for estimating expected returns and risk metrics.

## 3.3    Explain the concept of marginal probability and its role in multivariate probability analysis.

Marginal probability is a fundamental concept in probability theory and plays a significant role in multivariate analysis.

Marginal probability refers to the probability of a single event or variable occurring independently of other events or variables. In other words, it is the probability of one event happening without any consideration of the other potential events that could occur.

In multivariate analysis, marginal probabilities are computed by summing the joint probabilities over one of the variables. For example, if we have a joint probability distribution for two random variables X and Y, we can compute the marginal probability of X by summing the joint probabilities over all possible values of Y.

The role of marginal probabilities in multivariate analysis is to enable us to examine the influence of a particular variable on the overall probability distribution without considering the effects of other variables. These probabilities provide us with valuable insights into the behavior of individual variables in the presence of other variables, which is useful for making decisions in finance, economics, and other fields.

Marginal probabilities are often used in portfolio optimization and risk management, where investors need to analyze the potential returns and risks associated with each individual asset in a portfolio. By computing the marginal probabilities, analysts can develop better investment strategies that minimize risks and maximize returns.

Overall, marginal probabilities are a crucial concept in probability theory and multivariate analysis. They provide useful information about the behavior of individual variables in the presence of other variables and are vital for making informed decisions in finance, risk management, and other fields that rely on probabilistic models.

# 3.4 What is the law of total probability, and how is it used in quantitative analysis?

The Law of Total Probability is a fundamental concept in probability theory, which asserts that for any event A and a partition of the sample space (i.e., a collection of mutually exclusive and exhaustive events), the probability of A can be expressed as the sum of the conditional probabilities of A given each partition event, weighted by their respective probabilities.

Mathematically, the Law of Total Probability states that if B1, B2, ..., Bn are mutually exclusive and exhaustive events (i.e., the union of all the Bi equals the entire sample space), then for any event A:

$$P(A) = \sum_{i=1}^{n} [P(A|B_i) * P(B_i)] \tag{3.2}$$

This formula is based on the intuitive idea that the probability of A happening is the sum of the probabilities of A happening under each possible condition or scenario defined by the partition events B1, B2, ..., Bn, weighted by the probability of each scenario.

The Law of Total Probability is used extensively in quantitative analysis, particularly in the fields of data science, machine learning, and finance. In finance, quantitative traders and investors use the law of total probability to model the probability of future market movements or events based on available information.

For example, suppose a quantitative trader wants to estimate the probability of a stock moving up or down tomorrow given the current market conditions. The trader can partition the sample space into possible scenarios, such as bullish, bearish, or neutral underlying market trends. Then, the probability of the stock moving up or down tomorrow can be computed as a weighted average of the conditional probabilities of these scenarios. Specifically, the trader can estimate the conditional probabilities of the stock moving up or down tomorrow given each scenario, and weight them by their respective probabilities.

The Law of Total Probability provides a powerful tool for quantitative analysts to reason about complex probabilistic systems and make informed decisions based on data and statistical analysis.

## 3.5    Describe the Poisson distribution and its applications in quantitative finance.

The Poisson distribution is a mathematical concept that is widely used in probability theory and quantitative finance. It is a discrete probability distribution that describes the probability of a certain number of events occurring in a fixed time interval, given the average rate at which the events occur. The Poisson distribution is commonly used to model rare events such as defaults, defaults on bonds, and extreme market moves.

In the Poisson distribution, the rate parameter $\lambda$ represents the average number of events that occur per unit of time. The probability of observing a certain number (k) of events in a fixed time period is given by the Poisson distribution formula:

$$P(k) = (\lambda^k * e^{-\lambda})/k! \qquad\qquad (3.3)$$

where e is the natural logarithmic constant (approximately 2.71828), factorial (k!) represents the product of all positive integers from 1 to k, and $\lambda$ is the rate parameter.

The Poisson distribution is used in quantitative finance in a number of ways. One common application is in modeling the frequency of defaults on bonds or other debt securities. Financial institutions use the Poisson distribution to estimate the likelihood of multiple defaults occurring simultaneously, which can help them manage their credit risk.

Another common application of the Poisson distribution is in modeling market volatility. Financial traders often use the Poisson distribution to model and predict the frequency of extreme market moves, such as sudden drops or spikes in prices. This can help them make more informed trading decisions and manage their risk exposure.

In summary, the Poisson distribution is a powerful mathematical concept with applications in quantitative finance. It is commonly used to model rare events, and its use in modeling credit risk and market volatility helps financial institutions and traders make more informed decisions.

## 3.6 Explain the concept of hypothesis testing and its significance in quantitative analysis.

Hypothesis testing is a statistical method used in quantitative analysis to determine whether a specific assumption or hypothesis about a population or dataset is true or false. The hypothesis is typically in the form of a claim or statement about a population parameter (such as a mean or proportion).

The basic process of hypothesis testing involves the following steps:

1. Formulating a null hypothesis (H0) and an alternative hypothesis (Ha)

2. Selecting an appropriate statistical test and significance level ($\alpha$)

3. Collecting data and calculating a test statistic

4. Comparing the test statistic to a critical value to determine statistical significance

5. Drawing a conclusion based on the results of the analysis

The null hypothesis is the assumption that there is no difference or relationship between the variables being studied. The alternative hypothesis is the opposite of the null hypothesis, and it represents the possibility that there is a difference or relationship between the variables.

The significance level, alpha, is the probability of making a type I error, which is rejecting the null hypothesis when it is actually true.

It is typically set at 0.05 or 0.01, depending on the level of confidence desired.

The test statistic is a numerical value that measures the difference between the observed data and what would be expected if the null hypothesis were true. It is used to determine whether the null hypothesis should be rejected or not.

The critical value is a cutoff value that separates the region of rejection from the region of acceptance based on the significance level and degrees of freedom.

Hypothesis testing is significant in quantitative analysis because it allows researchers and analysts to make data-driven decisions and draw conclusions based on statistical evidence. By testing hypotheses and determining the probability of different outcomes, it is possible to measure the level of uncertainty in a given situation and make more informed decisions.

For example, in finance, a trader might use hypothesis testing to determine whether a particular trading strategy is statistically significant and likely to generate positive returns. By testing the hypothesis, the trader can assess the risk of the strategy and make more informed decisions about whether to invest in it or not.

## 3.7    What is the difference between Type I and Type II errors in hypothesis testing?

During hypothesis testing, we make an assumption about a population based on a sample. We then test this assumption using statistical methods. However, there are chances that our test results may not always be precise, and we might encounter errors in our hypothesis testing. These errors are broadly classified into two types, namely Type I and Type II errors.

Type I error:

A type I error is also known as a false positive error. This error occurs when we reject the null hypothesis when it is actually true. In other words, we conclude that there is a statistically significant difference between two groups when there is no difference. Therefore, there is a high chance of making a type I error, and we use the level of significance (alpha) to control this risk.

For example, lets say a pharmaceutical company has developed a new drug for a particular disease. The null hypothesis is that the drug has no effect, and the alternative hypothesis is that the drug has a beneficial impact. If the company rejects the null hypothesis and concludes that the drug is effective based on the sample data, but in reality, the drug has no effect, this will be a type I error.

Type II error:

A type II error is also known as a false-negative error. This error occurs when we fail to reject the null hypothesis when it is actually false. In other words, we conclude that there is no statistically significant difference between two groups when there is a difference.

For example, lets say a company wants to test a product's durability under different conditions. The null hypothesis is that the durability is the same for all conditions, and the alternative hypothesis is that durability differs with various circumstances. If the company fails to reject the null hypothesis and concludes that the product's durability is the same under all conditions, but in reality, the product's durability differs, this will be a type II error.

In quantitative trading and investment, hypothesis testing is a crucial tool to evaluate investment strategies and to estimate the expected returns. A type I error can lead to a false-positive indication of a strategys success, leading to overconfidence and resulting in poor investment decisions. A type II error can cause a false-negative indication of a strategys performance, leading to underestimation of strategy, and missing out on profitable trading opportunities. Therefore, it is critical to understand the differences between type I and type II errors and control these errors to ensure the accuracy of the hypothesis test results.

## 3.8    Describe the t-distribution and its use in hypothesis testing and confidence intervals.

The t-distribution is a probability distribution that is similar to the standard normal distribution, but it is used when the sample size is small, and the population standard deviation is unknown. The t-distribution is symmetrical and bell-shaped, with a mean of 0 and a standard deviation greater than 1.

The t-distribution is commonly used in hypothesis testing and confidence intervals. In hypothesis testing, it is used to determine if there is a significant difference between two sample means or if a sample mean is significantly different from a hypothesized population mean. The t-distribution is used because the sample size is often small, and the population standard deviation is unknown.

When conducting a hypothesis test, the t-distribution is used to find the p-value, which is the probability of obtaining a test statistic as extreme or more extreme than the observed test statistic, assuming the null hypothesis is true. If the p-value is less than the level of significance, then the null hypothesis is rejected.

In confidence intervals, the t-distribution is used to calculate the margin of error. The margin of error is the range of values that a population parameter is likely to fall within given the sample data. The t-distribution is used because it accounts for the uncertainty associated with estimating the population standard deviation from the sample data.

For example, suppose we want to conduct a hypothesis test to determine if the mean weight of a sample of 50 students is significantly different from the population mean weight of 140 pounds. If we assume that the population standard deviation is unknown, we would use the t-distribution to find the p-value. If the p-value is less than the level of significance, we would reject the null hypothesis and conclude that the mean weight of the sample is statistically different from the population mean weight.

In another example, suppose we want to construct a confidence interval for the mean height of a population of 100 individuals. If we have a small sample size, say 20, and the population standard deviation is unknown, we would use the t-distribution to calculate the margin of error and construct the confidence interval. The t-distribution allows us to estimate the standard error and adjust for the uncertainty associated with estimating the population standard deviation.

## 3.9   Explain the concept of a p-value and its role in hypothesis testing.

In hypothesis testing, we use statistical tests to make inferences about a population based on sample data. The p-value is a commonly used concept in hypothesis testing that helps us determine whether the observed effect in our sample is statistically significant or simply due to chance.

The p-value is defined as the probability of obtaining a test statistic as extreme as or more extreme than the one observed in our sample, assuming that the null hypothesis is true. The null hypothesis is the idea that there is no difference or relationship between the variables being studied, and any observed differences or relationships are due to chance.

The smaller the p-value, the less likely it is for the observed effect to be due to chance. Typically, a p-value of less than 0.05 (or 5%) is considered statistically significant, which means that we reject the null hypothesis and accept the alternative hypothesis, which states that there is a relationship or difference between the variables being studied.

For example, suppose we want to test whether a new drug is effective in reducing blood pressure compared to a placebo. We would randomly assign participants to either the new drug or placebo group, record their blood pressure readings, and calculate the mean difference between the two groups. We would then use a statistical test, such as a t-test, to determine the p-value.

If the p-value is less than 0.05, we can conclude that there is a statistically significant difference in blood pressure between the two groups, and the new drug is effective. However, if the p-value is greater than

0.05, we cannot reject the null hypothesis, and we cannot conclude that the new drug is effective.

In summary, the p-value is a critical concept in hypothesis testing that helps us determine the likelihood that our observed effect is due to chance. A small p-value indicates strong evidence against the null hypothesis and provides support for the alternative hypothesis.

## 3.10    Describe the Chi-squared distribution and its applications in quantitative finance.

The Chi-squared distribution, denoted by $\chi^2$ distribution, is a continuous probability distribution that arises from summing the squares of independent standard normal random variables. It is characterized by a positive parameter known as the degrees of freedom. The degrees of freedom determine the shape of the distribution and affect its mean and variance.

The Chi-squared distribution is commonly used in statistical inference, hypothesis testing, and model selection. In quantitative finance, the Chi-squared distribution is often used to test the goodness of fit of a model or to measure the amount of variability in a sample of data.

One common application of the Chi-squared distribution in quantitative finance is in the evaluation of the performance of an investment portfolio. When evaluating the performance of a portfolio, it is important to determine whether the returns are consistent with a normal distribution or whether there is excessive variability. By applying the Chi-squared goodness-of-fit test, analysts can determine if the portfolio follows a normal distribution or if some other distribution should be used.

Another application of the Chi-squared distribution in finance is in the construction of Value at Risk (VaR) models. VaR is a statistical measure of the potential loss in the value of an investment portfolio over a given time period. The Chi-squared distribution can be used to estimate the value of VaR by providing a measure of the variability in returns.

Finally, the Chi-squared distribution is also used in the assessment of the fit of asset pricing models. The Capital Asset Pricing Model (CAPM), for example, assumes that the returns on a portfolio are normally distributed. By using the Chi-squared distribution, analysts can test whether the observed returns deviate significantly from the normal distribution, indicating that the CAPM model may not be an appropriate model for asset pricing.

In summary, the Chi-squared distribution is a versatile tool for assessing the fit of statistical models, evaluating investment portfolio performance, and constructing VaR models in quantitative finance.

## 3.11  Explain the concept of statistical power and its importance in hypothesis testing.

Statistical power is the ability of a hypothesis test to detect a true effect or difference if it truly exists. In other words, it is the likelihood of a statistical test to reject a null hypothesis when the null hypothesis is false (i.e., when there is a real effect or difference to be detected). Statistical power is an essential concept in hypothesis testing because it helps determine whether a study is capable of detecting the effect that it intends to detect.

In hypothesis testing, we usually set a threshold for the level of significance (p), which is the probability of rejecting a true null hypothesis. For example, we may set $p = 0.05$, meaning that we're willing to accept a 5% chance of rejecting a true null hypothesis. If the power of the test is high, it means that the test is less likely to make a type II error (failing to reject a false null hypothesis), and more likely to detect a true effect or difference.

Power is affected by several factors, including the effect size, sample size, level of significance, and the variability of the data. A larger sample size or a smaller level of significance can increase the power of a test. Conversely, a smaller effect size or greater variability can decrease the power.

For example, imagine that we want to test whether a new drug can lower blood pressure. We set up our null hypothesis that the drug has no effect on blood pressure. If we conduct a study with a low statistical power, we might fail to reject the null hypothesis even if the drug actually does lower blood pressure. This could be due to a small sample size, high variability in the data, or a low significance level. On the other hand, if we conduct a study with high statistical power, we're more likely to detect a true effect if it exists, making it more likely that we'll discover whether the drug is effective.

Therefore, having a high statistical power is important for ensuring that a hypothesis test accurately reflects the effect or difference being studied. It helps to avoid type II errors, which can result in a missed opportunity to discover an effective intervention or to make informed decisions based on the data.

## 3.12    How do you apply the bootstrap method in quantitative analysis?

The bootstrap method is a powerful statistical tool that can be used to estimate the characteristics of a population from a single sample. In quantitative finance, the bootstrap method is often used in the analysis of financial time series data, where it is important to estimate the parameters of a model or the distribution of a variable.

The bootstrap method involves creating a large number of resamples from the original sample, and then using these resamples to estimate the characteristics of interest. The resamples are created by randomly selecting data points from the original sample with replacement, which means that some data points may be selected multiple times while other may not be selected at all.

There are several steps involved in applying the bootstrap method in quantitative analysis, including:

1. Collecting the original data: The first step in applying the bootstrap method is to collect the original data that will be used to estimate the characteristics of interest. This data may come from financial markets, economic indicators, or other sources.

2. Creating the resamples: Once the original data has been collected, the resamples can be created by randomly selecting data points from the original sample with replacement. The number of resamples created is typically very large, often in the thousands or tens of thousands.

3. Estimating the characteristics of interest: After the resamples have been created, the characteristics of interest can be estimated using these resamples. For example, if the goal is to estimate the mean return of a particular security, then the mean return can be calculated for each of the resamples, and the distribution of these means can be used to estimate the true mean return.

4. Evaluating the results: Finally, it is important to evaluate the results of the bootstrap analysis to determine their accuracy and reliability. This may involve comparing the estimated characteristics to those obtained from other methods, or testing the sensitivity of the results to different assumptions and parameters.

One example of the use of the bootstrap method in quantitative finance is in the estimation of VaR (Value at Risk). VaR is a measure of the potential loss a portfolio or investment may experience over a given time period. The bootstrap method can be used to estimate VaR by creating resamples from the historical return data of the portfolio, and then using these resamples to estimate the distribution of potential losses. This can help investors assess the risk associated with their investments, and make informed decisions about asset allocation and risk management.

## 3.13  What are copulas and how are they used in modeling multivariate dependence?

Copulas are mathematical functions that allow for modeling and quantifying the dependence structure between multiple random variables. They are used extensively in finance, risk management, and insurance to model multivariate dependence, such as how the prices of different assets are related to each other.

The key property of a copula is that it provides a way to separate the marginal distributions of the individual variables from their joint dependence structure. In other words, copulas allow us to model the correlation between variables without having to specify the shape or form of their individual distributions.

To get a sense of how this works, consider two stocks A and B, both of which can take on a range of values. Suppose we want to model how the two stocks are jointly distributed, that is, how likely different combinations of values are for these two stocks. One approach would be to estimate the marginal distributions of the two stocks separately, and then estimate their joint distribution assuming some particular correlation between them. However, this approach has several limitations, notably the inability to model arbitrary dependence structures.

Copulas overcome these limitations by providing a way to model the joint distribution directly, without having to specify the marginal distributions. The basic idea is to map the marginal distributions to a uniform distribution, which can then be combined using the copula function to create the joint distribution.

For example, consider a bivariate Gaussian copula, which is a popular way of modeling multivariate dependence. The first step is to transform the marginal distributions of the stocks to a standard normal distribution. Then, the dependency structure between the two stocks can be modeled using the correlation parameter of the Gaussian copula.

Once the joint distribution has been modeled using the copula, we can use standard techniques to calculate probabilities, such as the probability of a joint extreme event or the expected value of a portfolio of stocks.

Overall, copulas are a powerful tool for modeling multivariate dependence in a flexible and intuitive way. By separating the marginal distributions from the joint dependence structure, they provide a non-parametric approach that can accommodate complex dependence structures and capture tail dependence that are otherwise difficult to model.

# 3.14 Explain the difference between parametric and non-parametric statistical methods.

Parametric and non-parametric statistical methods are two different approaches to analyzing data in statistics and probability theory.

Parametric statistics assumes that the data follows a specific probability distribution, such as the normal distribution or the binomial distribution. This means that the parameters of the distribution can be estimated using the sample data, and the statistical tests can be based on these parameter estimates. For example, if we assume that a set of data follows a normal distribution, we can use the mean and standard deviation of the data to estimate the parameters of the normal distribution, and then use these parameter estimates to conduct hypothesis tests or confidence intervals.

On the other hand, non-parametric statistics does not make any assumptions about the underlying distribution of the data. Instead, it uses statistical methods that are based on ranking or order statistics of the sample data, such as the median or Wilcoxon signed-rank test. Non-parametric methods are often used when the data does not follow a normal distribution, or when the sample size is small and the normal distribution assumption may not be valid.

In general, parametric methods are more powerful and efficient when the assumptions are met, while non-parametric methods are more robust and flexible when the assumptions are not met. For example, if we have a large sample of data that we believe follows a normal distribution, we may choose to use a parametric method such as a t-test or ANOVA. However, if we have a small sample or suspect that the data does not follow a normal distribution, we may choose to use a non-parametric method such as the Wilcoxon rank-sum test or Kruskal-Wallis test.

In quantitative trading/investment, parametric methods are commonly used to model stock prices or financial returns, such as using the normal distribution to model daily stock returns. Non-parametric methods are sometimes used to detect patterns in financial time series data, such as using the Kolmogorov-Smirnov test to test for the presence of trends or cycles.

## 3.15   Describe the concept of information criteria, such as AIC and BIC, and their use in model selection.

Information criteria are statistical measures that are employed in selecting the best model from a set of competing models. These statistical measures take into account both the goodness of fit of the model and its complexity, ensuring that the selected model is not overfit or underfit.

Akaike Information Criterion (AIC) and Bayesian Information Criterion (BIC) are two popular information criteria used for model selection.

AIC is based on the principle that a good model should minimize the information loss during the prediction process. It measures the quality of a model through the difference between the maximum likelihood of the model and the number of model parameters. AIC is formulated as follows:

$$AIC = -2log(L) + 2k \tag{3.4}$$

where L is the likelihood of the model and k is the number of parameters in the model.

Unlike AIC, BIC takes into account the complexity of the model when computing the likelihood. It assumes that the true model is among the set of candidate models and assigns a higher penalty to models with more parameters. BIC is formulated as follows:

$$BIC = -2log(L) + log(n)k \tag{3.5}$$

where n is the sample size and the other terms are the same as in AIC.

The main advantage of using information criteria in model selection is that they balance the trade-off between model complexity and goodness of fit. They provide a principled way of selecting the most parsimonious model that best captures the patterns in the data.

For example, consider a linear regression problem with multiple predictors. Suppose we have two models - one with three predictors and another with six predictors. To select the best model, we can compute the AIC and BIC scores for each model and choose the one with the lower value. The model with 3 predictors is likely to have a lesser AIC and BIC value, indicating that it may be the better choice as compared to the model with 6 predictors. This approach ensures that the selected model is not overfit and generalizes well to unseen data.

## 3.16 What is the concept of stationarity and its importance in time series analysis?

In time series analysis, stationarity is a fundamental concept that refers to the statistical properties of a process that remain unchanged over time. Specifically, a stationary time series is one where the mean, variance, and autocovariance structure remain constant over time. This means that the statistical properties of the data do not change over time, and therefore, the series exhibits predictable patterns that can be modeled and exploited in trading and investment decisions.

The importance of stationarity in time series analysis lies in the fact that it enables us to make more accurate forecasts and predictions. If a time series is stationary, we can develop models that capture its underlying trends, cycles, and seasonality, and use these models to forecast future values. On the other hand, if a time series is nonstationary, it means that it is subject to trends, seasonality, and other time-varying factors that make forecasting more difficult.

For example, let's consider the S&P 500 index, which is a widely used benchmark for the US stock market. If we plot the daily returns of the index over a long period of time, we can see that there are fluctuations and trends that vary over time, but overall, the series appears to be stationary. This suggests that the statistical properties of the returns, such as the mean and variance, are relatively stable over time, and that we can develop models that capture these properties to make predictions about future returns.

In contrast, let's consider a non-stationary time series such as the price of a commodity like oil. The price of oil is subject to many factors that can change over time, such as global supply and demand, geopolitical events, and weather patterns. As a result, the oil market is often volatile and hard to predict. This makes it more difficult to develop accurate models to forecast its future price movements.

In conclusion, stationarity is an essential concept in time series analysis because it allows us to model and forecast data accurately, which is critical in quantitative trading and investment. By identifying the stationary properties of a time series, we can develop models that capture its underlying patterns and use these models to make more informed investment decisions.

## 3.17   Explain the difference between a white noise process and an autoregressive (AR) process.

In probability theory and time series analysis, a stochastic process is called a white noise process if it is a sequence of random variables that are independent and identically distributed with a mean of zero and a constant variance. In other words, a white noise process is a random sequence of uncorrelated observations with equal variance.

On the other hand, an autoregressive (AR) process is a stochastic process in which each observation is a linear combination of the previous observations and a random error term. The order of an AR process specifies how many previous observations the current observation is dependent upon. The simplest form of an AR process is the first-order AR process, or AR(1):

$$X_t = aX_{t-1} + e_t \tag{3.6}$$

where $X_t$ is the current observation, $X_{t-1}$ is the previous observation, a is a constant coefficient, and $e_t$ is a random error term.

The main difference between a white noise process and an AR process is that a white noise process has no dependence between the observations, whereas an AR process has dependence between the observations through the autoregressive relationship. Additionally, the autocorrelation function of a white noise process is zero for all lags except zero, while the autocorrelation function of an AR process decays exponentially as the lag increases.

For example, consider two time series: one generated from a white noise process and the other from an AR(1) process. The white noise process would have randomly fluctuating observations with no pattern, while the AR(1) process would have smoother observations that are dependent on the previous observation and the random error term. The autocorrelation function of the AR(1) process would also decay exponentially as the lag increases, while the autocorrelation function of the white noise process would be zero for all lags except zero.

In quantitative trading or investment, it is essential to understand the underlying process that generates the data to make the right predictions and investment decisions. A better understanding of white noise and AR processes can help in developing forecasting models and trading strategies that take into account the underlying dynamics of the time series data.

## 3.18 Describe the concept of a moving average (MA) process and its use in time series analysis.

A moving average (MA) is a commonly used time series method for smoothing data that involves calculating a series of averages of different subsets of the full data set. The MA process is a time series model that describes a sequence of observations, where each observation is defined as the weighted average of recent values in the series.

Specifically, a simple MA process of order q, denoted as MA(q), can be represented as:

$$Y_t = c + e_t + \mu_1 e_{t-1} + \mu_2 e_{t-2} + ... + \mu_q e_{t-q} \qquad (3.7)$$

where $Y_t$t is the observed value at time $t$, $c$ is a constant, $e_t$ is a white noise error term, and $\mu_1$, $\mu_2$, $\mu_q$ are coefficients that determine the weights assigned to the $q$ past error terms.

The idea behind the MA process is that the current value of the time series depends on the past white noise error terms, with the coefficients indicating the relative importance of the different past errors. The larger the value of q, the more past error terms included in the calculation of each smoothed value, resulting in a smoother series.

Moving average processes are commonly used in time series analysis for several purposes. One application is to filter out noise or random variations in the data, making patterns easier to identify. Another use is to develop a forecast or prediction of future data points based on past values. By fitting a MA model to historical data, we can use it to make predictions for future values.

For example, suppose we have monthly sales data for a retail store, including seasonal fluctuations and some random noise. We can use the MA process to smooth out the random noise and isolate the underlying seasonal pattern, which can help us understand the cyclical behavior of the business. We can also use it to forecast future sales by estimating the coefficients from past data and using them to predict future values. By incorporating the MA process into a larger model, such as an autoregressive moving average (ARMA) process, we can build more sophisticated time series models that capture both trend and seasonality.

## 3.19   Explain the GARCH model and its application in modeling financial time series.

The Generalized Autoregressive Conditional Heteroskedasticity (GARCH) model is a statistical model used to estimate and forecast the volatility of financial time series data, such as stock prices and exchange rates.

Volatility is a key measure of risk in financial markets, and understanding and predicting changes in volatility is essential for managing risk and making informed investment decisions.

The GARCH model is an extension of the autoregressive conditional heteroskedasticity (ARCH) model, which models the variance of a time series as a function of lagged squared residuals. The GARCH model adds another equation which includes lagged conditional variances to capture the persistence of volatility.

The GARCH model can be expressed as:

$$r_t = \mu + \epsilon_t \tag{3.8}$$

$$\epsilon_t = \sigma_t * z_t \tag{3.9}$$

$$\sigma_t^2 = \alpha_0 + \alpha_1 * \epsilon_{t-1}^2 + \beta * \sigma_{t-1}^2 \tag{3.10}$$

where $r_t$ is the return at time $t$, $\mu$ is the mean return, $\epsilon_t$ is the standardized residual at time t, $\sigma_t^2$ is the conditional variance at time $t$, $z_t$ is a standard normal random variable, and $\alpha_0$, $\alpha_1$, and $\beta$ are the model parameters.

The GARCH model allows for time-varying volatility by incorporating past information about the variance of the time series. The inclusion of lagged conditional variances captures the persistence of volatility, while the parameter estimates for and provide information about the short- and long-term impact of past shocks on future volatility.

The GARCH model has several applications in modeling financial time series data. One of its main uses is in risk management, where it is used to estimate and forecast future volatility for asset pricing models and portfolio management. It can also be used in trading strategies that rely on volatility, such as options trading or volatility forecasting.

For example, suppose an investor wants to hedge against potential losses in a particular stock. By estimating the future volatility of the stock using a GARCH model, the investor can determine the optimal size and timing of a protective put option or other hedging strategy.

Overall, the GARCH model is a powerful tool for modeling and forecasting financial time series data and has a range of applications in risk management and investment strategies.

## 3.20    How do you apply principal component analysis (PCA) in quantitative finance?

Principal Component Analysis (PCA) is a statistical technique that can be applied in quantitative finance to identify the underlying structure in datasets that contain a large number of interrelated variables. The technique seeks to reduce the dimensionality of the dataset by identifying a set of new variables, called principal components, that capture the most significant sources of variation in the original dataset.

PCA is commonly used in finance for tasks such as risk management, portfolio optimization, and asset pricing. Here is how PCA can be used in each of these areas:

1. Risk management: A common application of PCA in risk management is the estimation of value-at-risk (VaR). VaR is a measure of the potential loss in the value of a portfolio over a given time horizon, with a given level of confidence. PCA can be used to identify the principal components of a set of asset returns, representing the most significant sources of variation. These principal components can then be used to estimate the VaR of a portfolio, based on the historical distribution of these components.

2. Portfolio optimization: PCA can also be used to construct optimized portfolios. Rather than using the original asset returns to construct the portfolio, an investor can use the principal components identified by PCA, which capture the most significant sources of variation. This allows for the construction of more diversified portfolios, as a smaller number of principal components can be used to explain a larger proportion of the overall variability in the returns of the assets.

3. Asset pricing: PCA can be used to identify common factors that influence the returns of multiple assets. By analyzing the correlation structure of the returns, PCA can identify the principal components that capture the common sources of variation in the returns. These

principal components can then be used to estimate the returns of individual assets, based on their exposure to the common factors.

For example, in the case of interest rate modeling, PCA can be used to extract the principal components from a large set of yield curve data. The components can then be used to estimate the parameters of a multivariate stochastic process to simulate the future interest rate scenario to build bond portfolios. By doing so, an investor can gain a deeper understanding of the underlying drivers of interest rate movements and make informed investment decisions.

In conclusion, PCA is a powerful technique in quantitative finance that can be applied to a wide range of problems. By identifying the underlying structure in large datasets, PCA can help investors to make better-informed decisions in tasks such as risk management, portfolio optimization, and asset pricing.

# Chapter 4

# Advanced

## 4.1 Explain the concept of risk-neutral probability and its relevance in option pricing.

In options pricing, risk-neutral probability is a concept that refers to the probability of an event occurring based on the assumption that the market is risk-neutral. A risk-neutral market is one in which the expected return on all assets is the risk-free rate. In such a market, the expected value of an option at expiration is the present value of the option's payoff discounted at the risk-free rate.

To calculate the risk-neutral probability of an event occurring, we use the risk-neutral valuation method. The steps involved in this method are as follows:

1. Determine the expected cash flows from the option at expiration under different market scenarios.

2. Discount the expected cash flows at the risk-free rate to obtain the option's present value under each scenario.

3. Calculate the probability of each scenario occurring under the risk-neutral assumption such that the expected value of the

option is equal to its present value.

For example, suppose you are considering a call option on a stock with a current price of $100 and a strike price of $110. Assume that the risk-free rate is 5%, and there are two possible outcomes at expiration: either the stock price will increase to $120 or decrease to $90. You can use the risk-neutral probability to calculate the option's fair price.

Using the risk-neutral valuation method, you can calculate the expected value of the call option at expiration under the two scenarios as follows:

- If the stock price increases to $120, the option is "in the money," and its payoff is $10 ($120 - $110). The option's expected value is therefore $10. Since the risk-free rate is 5%, the option's present value under this scenario is $9.52 [= $10 / (1 + 5%)].

- If the stock price decreases to $90, the option is "out of the money," and its payoff is $0. The option's expected value is therefore $0, and its present value under this scenario is also $0.

Next, you need to calculate the risk-neutral probabilities of the two scenarios such that the expected value of the option is equal to its fair price. Let P be the probability of the stock price increasing to $120, and (1-P) be the probability of the stock price decreasing to $90. Then, the fair price of the call option is:

$9.52P + $0(1-P) = $9.52P

Since the expected value of the call option is $9.52, we can set this equal to its fair price and solve for P:

$9.52 = $9.52P P = 1

Therefore, the risk-neutral probability of the stock price increasing to $120 is 1, and the probability of the stock price decreasing to $90 is 0. This implies that the market is assuming a risk-neutral stance, and the fair price of the call option is $9.52.

In summary, risk-neutral probability is a key concept in option pricing as it allows one to calculate the fair price of an option based on the assumption of a risk-neutral market. This approach has become widely used in quantitative finance because it provides a powerful tool for valuing complex financial instruments.

## 4.2 Describe the Black-Scholes-Merton option pricing model and its key assumptions.

The Black-Scholes-Merton (BSM) model is a widely used quantitative tool for pricing European-style options on underlying assets, such as stocks, indices, currencies, and commodities. The BSM model is based on the concept of a risk-neutral probability, which allows for the derivation of a mathematical formula for the fair value of an option. The BSM model assumes that the underlying asset follows a lognormal distribution and that the option can be perfectly replicated by a combination of the underlying asset and a riskless bond. The key assumptions of the BSM model are:

1. The underlying asset price follows a lognormal distribution: The BSM model assumes that the underlying asset price follows a random walk process and that the log returns of the asset are normally distributed. This assumption has been criticized for being too simplistic, as it does not account for the fat tails and skewness observed in many asset price distributions.

2. The option is European-style: The BSM model is designed to price European-style options, which can only be exercised at expiration. This assumption allows for the derivation of a closed-form solution for the option price, which is not possible for American-style options that can be exercised at any time prior to expiration.

3. No arbitrage opportunities exist: The BSM model assumes that there are no arbitrage opportunities in the market, meaning that it is not possible to make riskless profits by taking advantage of price discrepancies. This assumption is used to derive the risk-neutral probability measure.

4. The risk-free rate and volatility are constant: The BSM model assumes that the risk-free interest rate and the volatility of the underlying asset are constant over the life of the option. This assumption may not hold in practice, as interest rates and volatility can change

over time.

5. No dividends are paid on the underlying asset: The BSM model assumes that the underlying asset does not pay any dividends during the life of the option. This assumption can be relaxed by adjusting the model to account for dividends.

Overall, the BSM model is a powerful tool for pricing options and is widely used by practitioners in the financial industry. However, it is important to understand the key assumptions of the model and their implications for option pricing in practice.

## 4.3 How do you apply stochastic calculus in quantitative finance? Provide an example.

Stochastic calculus is a mathematical tool used to model and analyze systems that involve probabilistic or random components. In quantitative finance, stochastic calculus is used to model the movements and dynamics of financial instruments over time, incorporating factors such as uncertainty, volatility, and risk.

One application of stochastic calculus in quantitative finance is the modeling of stock prices using the geometric Brownian motion (GBM) model. This model assumes that stock prices follow a random walk with a drift and a volatility that are both stochastic processes. The drift represents the expected return of the stock and is often modeled as the risk-free rate, while the volatility represents the uncertainty or randomness in the stock price movements.

For example, suppose we want to model the stock price of a company over a period of time using the GBM model. We can write the stock price as follows:

$$S(t) = S(0)exp((r - 0.5 * \sigma^2)t + \sigma * W(t)) \qquad (4.1)$$

where S(t) is the stock price at time t, S(0) is the initial stock price, r is the risk-free rate, $\sigma$ is the volatility of the stock price, t is time, and W(t) is a Brownian motion or Wiener process.

Using this model, we can simulate the future stock price movements of the company by generating a sequence of random numbers that follow a normal distribution with mean 0 and variance 1. We can then plug these random numbers into the GBM formula to calculate the corresponding stock prices over time.

Stochastic calculus also plays a crucial role in the development of derivative pricing models, such as the Black-Scholes model. The Black-Scholes model allows us to price options by assuming that the underlying asset follows a GBM process and that the option payoff is a function of this underlying asset. Stochastic calculus is used to derive the partial differential equation (PDE) that governs the behavior of the option price over time, using techniques such as Ito's lemma and the Feynman-Kac theorem.

Overall, stochastic calculus serves as a powerful tool for modeling the complex and uncertain dynamics of financial markets and instruments, allowing us to make informed investment and trading decisions.

## 4.4 Explain the concept of a Brownian motion and its role in stochastic processes.

Brownian motion is a type of stochastic process that describes the random movement of particles in a fluid or gas. The term "Brownian" comes from the name of the biologist Robert Brown, who observed this phenomenon in 1827 when he was looking at pollen grains under a microscope. He noticed that the particles were moving in a random way, even though there was no visible external force acting on them.

In the context of finance and investment, Brownian motion is used to model the random price fluctuations of securities and other financial instruments. This is based on the assumption that the stock or asset price follows a random and continuous path, which is driven by an underlying source of randomness.

The mathematical properties of Brownian motion are well understood and can be described by stochastic calculus. The most important property of Brownian motion is its "independence of increments": the movement of the particle at any given time is not influenced by its previous path or position. This makes it an ideal model for modeling unpredictable or random events, such as stock prices or fluctuations in interest rates.

In quantitative trading, Brownian motion is often used as a key building block in the creation of mathematical models and trading strategies. For example, the Black-Scholes option pricing model is based on Brownian motion and it is widely used by traders to price options and other derivative securities.

Overall, Brownian motion is an important concept in probability theory and quantitative finance, and it plays a crucial role in modeling and predicting the behavior of complex systems where randomness is a key factor.

## 4.5    Describe the geometric Brownian motion model and its application in option pricing.

The geometric Brownian motion (GBM) model is a widely used stochastic process in finance that describes the evolution of a variable over time, such as the price of a stock or other financial instrument. The GBM model is particularly useful for modeling stock prices, where the assumption is made that the stock price changes randomly over time and that the rate of change is proportional to the stock price itself.

Mathematically, the GBM model is expressed as:

$$dS(t) = \mu S(t)dt + \sigma S(t)dW(t) \qquad (4.2)$$

where S(t) is the stock price at time t, $\mu$ is the expected return of the stock, $\sigma$ is the volatility of the stock, dt is a small time increment, dW(t) is a Wiener process (i.e., a continuous-time stochastic process that models random fluctuations), and the differential equation dS(t) represents the instantaneous rate of change of the stock price.

In the context of option pricing, the GBM model is used to calculate the expected future value of an underlying asset, such as a stock, that an option is based on. This expected future value is known as the "forward price", and is used to determine the value of the option at expiration.

The Black-Scholes model is a widely used pricing model for European call and put options that uses the GBM model as its underlying assumption. The Black-Scholes model assumes that the stock prices follow a GBM process and that the risk-free rate is constant. Using this model, the value of a call option can be calculated using the formula:

$$C = S(t)N(d1) - Ke^{(-r(T-t))}N(d2) \qquad (4.3)$$

where C is the price of the call option, S(t) is the current spot price of the underlying asset, N(.) is the standard normal cumulative distribution function, K is the strike price of the option, r is the risk-free interest rate, T is the time until expiration, t is the current time, and d1 and d2 are calculated using the Black-Scholes formula:

$$d1 = (ln(S(t)/K) + (r + \sigma^2/2)(T - t))/(\sigma(T - t)^{1/2}) \qquad (4.4)$$
$$d2 = d1 - \sigma(T - t)^{1/2} \qquad (4.5)$$

In summary, the GBM model is a widely used stochastic process in finance that describes the evolution of a variable over time, such as stock prices, and is used extensively in option pricing models like the Black-Scholes model.

# 4.6 What is the binomial option pricing model and how does it work?

The binomial option pricing model is a popular method used in quantitative finance for pricing options. It is based on the assumption that the underlying asset follows a binomial distribution, where the stock price can either go up or down with a certain probability over a given time period. The model allows us to value options by determining the probabilities of different outcomes and calculating the expected value of the option at expiration.

The basic steps in the binomial option pricing model are as follows:

1. Divide the time period into equal intervals, such as months, weeks, days, or even hours.

2. Calculate the probability of the underlying asset going up or down for each time interval. This probability is often estimated using historical data or implied volatility.

3. Calculate the up and down movements of the stock price for each time interval, which will depend on the size of the up and down movements as well as the time interval.

4. Construct a binomial tree that shows all possible price paths of the underlying asset over the relevant time period.

5. Work out the payoff of the option at expiration for each possible price path on the tree.

6. Move backwards along the tree to calculate the option price at each earlier time point, taking the expected value of the option at each node. This expected value is calculated by weighting the possible future values by their associated probabilities of occurring.

For example, suppose we have a stock that is currently trading at $100, and we want to price a call option with a strike price of $110 that expires in 3 months. We assume that the stock price can either go up by 10% or down by 10% with equal probability in each month. We construct a binomial tree as shown below:

```
                                110
                121                             99
        133.1           108.9           108.9           89.1
   146.41      119.79  119.79    97.29         119.79   97.29
```

```
S0 = $100                                    S3 = ?
```

Each node on the tree represents a possible price of the stock in 3 months, depending on whether it goes up or down in each of the three months. We can then work out the option payoff at expiration for each node, which is the maximum of zero and the difference between the stock price and the strike price. For example, at the top node where the stock goes up in each of the three months, the option payoff is $0, since the stock price is below the strike price of $110.

We can then move backwards along the tree to calculate the option price at each of the earlier nodes, taking the expected value of the option at each node. For example, at the nodes where the stock price is $119.79 and $97.29, the expected value of the option is:

Expected value of option $= [0.5$ x (Value of option if stock goes up) $+ 0.5$ x (Value of option if stock goes down)]

```
= [0.5 x (119.79 - 110) + 0.5 x (0)]
= $4.895
```

We continue this process all the way back to the initial node, where the expected value of the option is the final option price. In this case, we find that the expected value of the option is $7.778. Therefore, the fair price of the call option using the binomial option pricing model is $7.778.

## 4.7   Explain the concept of delta hedging and its importance in managing option risk.

Delta hedging is a technique used in options trading to manage the risk associated with price movements of the underlying asset. The concept of delta refers to the sensitivity of the option price to changes in the price of the underlying asset.

Delta is a numerical value that ranges from 0 to 1 for a call option and from -1 to 0 for a put option. A delta of 0.5 means that for every

$1 increase in the price of the underlying asset, the option price will increase by 50 cents.

Delta hedging involves taking an offsetting position in the underlying asset to neutralize the delta value of an options portfolio. This means that if the delta of an options portfolio is positive, a short position is taken in the underlying asset to offset the expected price increase, and vice versa if the delta is negative.

For example, if an investor owns a call option with a delta of 0.5 and the price of the underlying asset increases by $1, the value of the call option will increase by $0.50. To neutralize this exposure, the investor can take a short position in the underlying asset with a value of $0.50. This allows the investor to lock in a profit regardless of the direction of the underlying asset price.

Delta hedging is important in managing option risk because it helps to reduce the impact of price movements of the underlying asset on the value of the options portfolio. By neutralizing the delta exposure, investors can eliminate the risk associated with price movements and focus on other factors such as time decay and volatility.

Overall, delta hedging is a crucial technique in options trading that allows investors to manage the risks associated with price movements of the underlying asset, and enables them to maximize their profits while minimizing potential losses.

## 4.8    Describe the Greeks in option pricing and their significance in risk management.

The Greeks are a set of mathematical measures used to analyze and manage risk in option pricing. These measures quantify the sensitivity of an options price to changes in underlying factors, such as the price of the underlying asset, volatility, time to expiration, and interest rates. The most commonly used Greeks are delta, gamma, theta, vega, and rho.

Delta measures the change in an options price for a $1 change in the underlying asset price. For a call option, delta is positive, indicating the option price increases as the asset price rises. For a put option, delta is negative, indicating the option price decreases as the asset price increases.

Gamma measures the rate of change of delta as the underlying asset price changes. It shows how much delta will change for a $1 change in the underlying asset price. Gamma is highest for at-the-money options and decreases as the option moves further in or out of the money.

Theta measures the rate of change in an options price with time. It shows how much the option price will decrease (for long positions) or increase (for short positions) as the time to expiration approaches. Theta is highest for at-the-money options and decreases as the option moves further in or out of the money.

Vega measures the sensitivity of an options price to changes in volatility. It shows how much the option price will increase (for long positions) or decrease (for short positions) for a 1% increase in implied volatility.

Rho measures the sensitivity of an options price to changes in interest rates. It shows how much the option price will increase (for long positions) or decrease (for short positions) for a 1% increase in interest rates.

The Greeks are significant in risk management as they help traders and investors understand the risks associated with their options positions and make informed trading decisions. By monitoring and adjusting their exposure to the various Greeks, traders can manage their portfolios sensitivity to market changes, limit potential losses, and optimize their returns. For example, a trader may use delta hedging to offset the delta risk of an option position, or may adjust their vega exposure to benefit from changes in implied volatility.

# 4.9    Explain the concept of implied volatility and its importance in option pricing.

Implied volatility (IV) is a measure of the expected volatility of the underlying asset that is implied by the current prices of option contracts. It is an indicator of the market's expectation of how much the underlying asset is expected to move up or down in the future.

In options trading, volatility is a crucial component in determining the fair value of an option. The higher the volatility, the more expensive the options premium will be, while low volatility will result in cheaper options. This is because an increase in volatility increases the likelihood of the option expiring in-the-money, and vice versa.

Implied volatility is important in option pricing because it indicates the market's opinion of how volatile the underlying asset will be in the future, which affects the option price. Options traders can use implied volatility to assess the potential risk and reward of an options trade. A high IV indicates that the market is expecting a significant price movement in the underlying stock or asset, making it riskier but potentially more profitable, while a low IV indicates that the market is expecting relatively little movement, making it less risky but less potentially profitable.

For example, suppose a stock is currently trading at $100 per share, and a call option with a strike price of $110 and expiration in 30 days is trading at $3 per contract. The implied volatility for the option is 20%. If the implied volatility increases to 30%, the premium of the option will increase to reflect the higher expected price movement of the underlying asset. Conversely, if the implied volatility decreases to 10%, the premium would decrease to reflect the lower expected price movement of the underlying asset.

In conclusion, understanding and analyzing implied volatility is crucial for option traders as it provides valuable insight into the market's expectations for the underlying stock or asset, and helps them gauge the potential risks and rewards of an options trade.

# 4.10   Describe the Heston model for stochastic volatility and its applications in quantitative finance.

The Heston model is a popular stochastic volatility model used in quantitative finance to price derivatives and calculate risk metrics such as Value-at-Risk.  Introduced by Steven Heston in 1993, the model is widely used by practitioners due to its ability to generate volatility smile (a phenomenon where out-of-the-money options have higher implied volatility than in-the-money options) and the tractability of its solution.

The Heston model describes the dynamics of the stock price (S) and the volatility (v) as two stochastic processes.  More specifically, it assumes that the instantaneous volatility v follows a mean-reverting square root process:

$$dv = \kappa(\theta - v)dt + \omega\sqrt{v}dW_v \qquad (4.6)$$

where is the mean-reversion speed parameter, is the long-term mean volatility, is the volatility of volatility parameter, and $W_v$ is a Brownian motion.

The Heston model also assumes that the stock price S follows a geometric Brownian motion with a drift term and a volatility that is a function of v:

$$dS = \mu Sdt + \sqrt{v}SdW_s \qquad (4.7)$$

where $W_s$ is another Brownian motion that is correlated with $W_v$.

The correlation between the two Brownian motions is given by the correlation coefficient .  This means that changes in the stock price are dependent on changes in volatility, and vice versa.

One of the main applications of the Heston model is to price derivatives, such as options.  The Heston model can be used to calculate the fair price of an option, given the market price of the underlying asset, the strike price of the option, the time to expiration, and other inputs such as interest rates and dividends.

The Heston model can also be used to calculate risk measures such as Value-at-Risk by simulating the stock price and volatility paths, and calculating the distribution of future returns. This allows traders and risk managers to estimate the potential losses that may occur under different market scenarios.

Overall, the Heston model is a powerful tool in quantitative finance due to its ability to generate realistic volatility surfaces and its utility in pricing derivatives and calculating risk measures.

## 4.11   What is jump diffusion modeling and its relevance in option pricing?

Jump diffusion modeling is a type of stochastic process that incorporates both continuous diffusion and occasional, sudden jumps in the underlying asset's price into the model. It is used to account for the market dynamics where the price of an asset experiences sudden jumps due to some unexpected event or news. These jumps cannot be adequately modeled with just a continuous stochastic process, such as a Brownian motion.

Jump diffusion modeling is widely used in the field of quantitative finance for modeling asset prices, interest rates, and credit risk. One of its most significant contributions to option pricing is the ability to incorporate not only the volatility of the underlying asset's price but also the likelihood of sudden jumps in the price.

In option pricing, jump diffusion models are used in the valuation of European options, as they can better capture the behavior of the underlying asset's price than traditional Black-Scholes models. The jump diffusion model considers the possibility of sudden jumps in the asset price by incorporating a Poisson process into the model, which affects the expected return of the underlying asset.

For example, suppose a stock has a jump in price due to an unexpected positive event, such as a successful drug trial for a pharmaceutical company. In that case, the jump diffusion model can correctly capture the stock's subsequent behavior, including both the increase in volatility and the upward momentum of the price.

In summary, jump diffusion modeling is an essential tool in quantitative finance to account for sudden jumps in the price of an asset that cannot be modeled by continuous diffusion. The incorporation of jumps into the model has proven to be valuable in accurately pricing options, especially in situations where sudden price movements are likely to occur.

## 4.12    Explain the concept of value at risk (VaR) and its importance in risk management.

Value at Risk (VaR) is a quantitative metric used to assess the potential loss that may be incurred due to adverse market movements within a specific time horizon, typically a day or a week. Its importance lies in its ability to provide a single summary measure of risk exposure that can be used by traders, portfolio managers, and risk managers to make informed decisions. VaR is typically expressed in monetary terms and represents the maximum potential loss that an investor could incur with a certain level of confidence (e.g., 99% or 95%).

VaR is calculated by looking at the historical returns of a portfolio or security, and determining the worst-case scenario losses based on a specified confidence level. For example, a portfolio with a 1-day 95% VaR of $1 million means that there is a 5% chance (confidence level) that the portfolio will lose more than $1 million in any given day.

VaR is an important risk management tool because it provides a way to quantify the potential risk of a portfolio or investment strategy. By knowing the VaR, investors can make informed decisions about how much risk they are willing to take on and adjust their positions accordingly. VaR can also be used by banks and other financial institutions to comply with regulatory requirements, such as the Basel Accords, which mandate minimum capital requirements based on VaR.

However, it is important to note that VaR is not a perfect measure of risk as it assumes that market returns follow a normal distribution, which is not always the case. VaR can also be affected by outliers or extreme market events that are not captured by historical data. Therefore, VaR should be used in conjunction with other risk metrics and stress testing to provide a more comprehensive view of risk exposure.

## 4.13  Describe the concept of conditional value at risk (CVaR) and how it differs from VaR.

Value at Risk (VaR) is a widely used risk measure in quantitative finance that measures the maximum loss a portfolio is expected to incur over a specified time horizon and confidence level. The VaR measure refers to the amount of money that a portfolio would stand to lose at a given level of confidence within a given time frame. VaR tells you how much you could lose, but it does not tell you anything about the magnitude of the loss that may exceed VaR.

Conditional Value at Risk (CVaR), on the other hand, is a more comprehensive risk measure that measures the expected value of all losses that exceed a given VaR threshold. In other words, it measures the average loss that an investor could experience beyond the VaR level, under the condition that the loss does exceed the VaR level.

CVaR is sometimes referred to as the expected tail loss or expected shortfall, as it measures the average size of the loss that occurs beyond the VaR threshold. CVaR considers all possible losses that may have occurred beyond the VaR level, and then calculates the average loss for those losses. It therefore provides a more complete picture of the worst-case losses that an investor may experience.

For example, let us suppose that you have invested into a portfolio of stocks and you have calculated the VaR of the portfolio to be $10 million, with a 95% confidence interval over a month. This means that there is a 5% chance that you will lose more than $10 million in a given month. Now, you are interested in knowing the expected size of loss in the event that the actual loss exceeds the VaR level. This is where CVaR comes into the picture. Let us suppose that the CVaR for the portfolio is $15 million. This means that if, hypothetically, you lose more than $10 million, the average size of your loss would be $15 million.

In summary, CVaR is a risk measure that provides a more comprehensive view of the downside risk associated with a portfolio than VaR alone, by taking into account the expected size of the loss beyond the VaR level.

## 4.14 How do you estimate portfolio risk using factor models, such as the CAPM and Fama-French models?

To estimate portfolio risk using factor models, such as the CAPM and Fama-French models, there are a few steps involved:

1. Define the factors: The first step is to define the factors that are relevant to the risk of the portfolio. For example, in the CAPM model, the single factor is the market risk premium (the difference between the return of the market and the risk-free rate). In the Fama-French model, there are three factors: market risk premium, size (small vs. large companies), and value/growth (value vs. growth companies).

2. Estimate factor exposures: Once the factors have been defined, the next step is to estimate the portfolio's factor exposures. This involves calculating the beta coefficients (or factor loadings) for each factor. Betas are a measure of how sensitive a stock or portfolio is to changes in the factor.

3. Calculate factor returns: The third step is to calculate the returns of each factor. This involves analyzing historical data to determine the returns of the market, small vs. large companies, and value vs. growth companies.

4. Calculate the expected return of the portfolio: With the factor exposures and factor returns calculated, you can then use them to estimate the expected return of the portfolio. This involves multiplying the beta coefficients for each factor by the expected return of that

factor and summing the products. The result is an estimate of the expected return of the portfolio based on the factor model.

5. Calculate portfolio risk: The final step is to calculate the portfolio's risk using the factor model. This involves calculating the portfolio's volatility by multiplying the beta of each factor by the standard deviation of that factor and summing the products. The result is an estimate of the portfolio's risk based on the factor model.

It's worth noting that factor models are just one way to estimate portfolio risk. Other methods include historical simulation, Monte Carlo simulation, and value at risk (VaR). However, factor models are widely used in quantitative finance and have been shown to be effective in estimating portfolio risk.

## 4.15   What is the concept of cointegration and its application in quantitative trading strategies?

Cointegration is a statistical concept that describes the long-term relationship between two or more non-stationary variables. In finance, cointegration is used to identify pairs or groups of assets that are likely to move together over time.

When two assets are cointegrated, it means that they share a common trend and their deviations from this trend are mean-reverting. In other words, if the two assets become too far apart from their common trend, they are likely to converge back towards it in the long run. This presents an opportunity for quantitative trading strategies that exploit this mean-reversion property.

One popular cointegration-based trading strategy is called pairs trading. In pairs trading, you identify two cointegrated assets and then look for periods when they diverge from each other. You then take a long position in the asset that has fallen more than it should have relative to the other and a short position in the asset that has risen more than it should have. As the assets revert back towards their common trend, the profits from the winning trade should exceed the losses from the losing trade, resulting in a profitable overall trade.

For example, let's say you identify two cointegrated stocks, A and B. You calculate the spread between the two stocks (i.e., A - B) and find that it is currently larger than usual. You believe this is due to a temporary shock that has affected only one of the stocks. You enter a long position in stock A and a short position in stock B. As the stocks revert back towards their common trend, the spread between them narrows, resulting in a profit for the trade.

Cointegration is also used in other quantitative trading strategies, such as mean-reversion and statistical arbitrage. However, it's important to note that cointegration is not a guarantee of profits and that these strategies come with their own set of risks, such as transaction costs and model risk. As with any trading strategy, thorough research and rigorous risk management are essential.

## 4.16   Describe the concept of mean reversion in financial markets and its use in trading strategies.

Mean reversion is a statistical concept that describes the tendency of a variable to return to its average value over time. In financial markets, mean reversion is often observed in asset prices, especially when prices have deviated significantly from their historical average. This phenomenon is often driven by investor behavior and irrational market reactions to news or events.

In equity markets, mean reversion is observed through the behavior of price-to-earnings (P/E) ratios. High P/E ratios typically indicate bullish sentiment and expectations for future earnings growth, while low P/E ratios can indicate bearish sentiment and undervalued stocks. Mean reversion suggests that stocks with high P/E ratios will eventually see those ratios decline, while undervalued stocks with low P/E ratios will see them rise over time.

Mean reversion can be utilized in trading strategies in a number of ways. One common strategy is to identify assets that have deviated significantly from their historical averages and take positions that expect them to revert to their mean value. This can involve buying undervalued assets or selling overvalued assets.

Another strategy involves trading the spread between two assets that exhibit mean reversion tendencies. For example, pairs trading involves buying the stock that has become undervalued relative to its historical average and simultaneously selling the stock that has become overvalued relative to its historical average. This type of strategy can be used to generate profit while limiting exposure to broader market conditions.

It's important to note that while mean reversion strategies can be profitable, they are not without risk. Extensive research and analysis of historical data is required to identify and monitor the assets that exhibit mean reversion tendencies. Additionally, market conditions and investor behavior can change quickly, making mean reversion trades vulnerable to sudden shifts in sentiment.

## 4.17   Explain the concept of pairs trading and its application in quantitative finance.

Pairs trading is a quantitative trading strategy that involves buying and selling two highly correlated financial instruments simultaneously in order to profit from the difference in their prices. The idea behind pairs trading is that when two stocks are highly correlated, they tend to move in the same direction most of the time. However, occasionally, one stock may deviate from the other due to temporary market inefficiencies or other factors, and pairs traders look to take advantage of these deviations.

To implement pairs trading, the trader first identifies a pair of stocks that are highly correlated, such as two tech giants Apple and Microsoft. The trader then calculates the ratio of the prices of the two stocks, and tracks the difference between the ratio and its historical average. When the ratio moves outside of certain thresholds or statistical measures, the trader takes a position accordingly. For example, if the ratio becomes too high and deviates from its usual range, the trader would sell Microsoft and buy Apple in hopes that the ratio will revert back to its mean, allowing them to profit from the difference in prices.

Another important aspect to consider in pairs trading is the selection of the appropriate time horizon. Since pairs traders are looking to profit from short-term deviations in prices, they typically hold positions for only a few days or weeks, and may need to make frequent adjustments to their positions as the market evolves.

Pairs trading is especially popular in quantitative finance due to its reliance on statistical analysis and historical data to identify opportunities. With the use of sophisticated statistical tools and algorithms, pairs traders can quickly identify pairs of stocks with high correlation and profitable trading thresholds. Additionally, since pairs trading involves buying and selling two stocks simultaneously, it is considered market-neutral, which reduces overall portfolio risk and helps to minimize exposure to broader market movements.

In summary, pairs trading is a quantitative trading strategy that involves buying and selling two highly correlated financial instruments in order to profit from temporary deviations in their prices. Its application in quantitative finance is widespread due to its reliance on statistical analysis and historical data, and its market-neutral nature that minimizes portfolio risk.

## 4.18   What is the concept of algorithmic trading and its importance in modern finance?

Algorithmic trading is a process of using computer programs to execute trade orders automatically, based on pre-defined set of rules and conditions. It is a subset of quantitative trading that relies on statistical models, mathematical algorithms, and data analysis to make trading decisions. Algorithmic trading is commonly used in equity, futures and foreign exchange markets, and is increasingly popular in fixed income and commodity markets as well.

Algorithmic trading has become a critical part of modern finance for several reasons:

1. Efficiency: Algorithmic trading enables market participants to conduct trades with high precision and speed, reducing or eliminating

the need for human intervention. Automated trading systems can evaluate market data and execute trades in fractions of a second, significantly faster than a human trader.

2. Cost reduction: Algorithmic trading can reduce trading costs for market participants. By automating the trading process, algorithmic trading can minimize the impact of execution costs such as slippage, bid-ask spreads, and commissions.

3. Data-driven decision making: Algorithmic trading relies on data analysis to make informed trading decisions, ensuring that trades are backed by robust statistical models and accurate market data.

4. Scalability: Algorithmic trading systems can be programmed to handle large volumes of trades simultaneously, making it easier for traders to manage a large portfolio of stocks or other securities.

Overall, the concept of algorithmic trading has revolutionized the way markets are traded, enabling market participants to use advanced technology and data analysis to make more informed trading decisions.

## 4.19   Explain the use of machine learning techniques, such as neural networks and decision trees, in quantitative finance.

Machine learning techniques, such as neural networks and decision trees, have found widespread applications in quantitative finance in recent years. These techniques have the ability to automatically learn and adapt to complex patterns in data, making them useful for a variety of tasks, including forecasting asset prices, risk management, portfolio optimization, and algorithmic trading.

Neural networks, for example, are a type of machine learning algorithm inspired by the structure of the human brain. They can be used to make predictions and classifications based on input data. In finance, they are often used for time series analysis to forecast future price movements. For example, recurrent neural networks (RNNs) can be used to model the temporal dependencies of financial time series data and make short-term predictions of intraday prices.

Another popular machine learning technique in finance is decision trees. These are tree-like models that split data into smaller and smaller subsets based on certain criteria. They can be used for classification or regression problems, such as predicting whether an asset will go up or down in price. Decision trees are often used in credit risk assessment, where they can be used to make loan approval decisions based on data such as credit scores, income, and employment history.

Other machine learning techniques that are commonly used in quantitative finance include support vector machines (SVMs), random forests, and clustering algorithms. These techniques can be used for a variety of tasks, such as asset allocation, risk management, and fraud detection.

However, it is important to note that while machine learning techniques can be powerful tools in finance, they should not be used blindly. Proper validation and testing of models is critical to ensure that they are effective and reliable. In addition, machine learning techniques should be used in conjunction with fundamental analysis and other quantitative methods to generate robust investment strategies.

## 4.20 Describe the concept of reinforcement learning and its applications in algorithmic trading strategies.

Reinforcement learning is a subfield of machine learning where an agent learns to make decisions based on trial-and-error experience. In reinforcement learning, an agent interacts with an environment and receives a reward signal based on its actions. The goal is to learn a policy that maximizes the expected cumulative reward over time.

In the context of algorithmic trading strategies, reinforcement learning can be applied to learn a trading policy that maximizes profit or minimizes risk. The environment could be a financial market, and the agent's actions could be buying or selling an asset. The reward signal could be the profit or loss made by the agent.

One of the key advantages of reinforcement learning in trading is that it can adapt to changing market conditions. For example, if the market undergoes a sudden and unexpected change, a reinforcement learning agent can quickly learn to adjust its trading policy based on the new conditions.

There are several applications of reinforcement learning in algorithmic trading. One example is using reinforcement learning to optimize the execution of trades. In this case, the agent learns to execute trades at the best possible time and price, taking into account factors such as transaction costs and market liquidity.

Another application is using reinforcement learning to develop trading strategies. In this case, the agent learns to make trading decisions based on a combination of technical indicators and market information. The advantage of using reinforcement learning for strategy development is that the agent can learn to exploit complex patterns in the data that may be difficult for human traders to identify.

Overall, reinforcement learning is a powerful approach to developing algorithmic trading strategies, especially in rapidly changing market conditions where traditional approaches may not work well. However, it is important to carefully validate and test any reinforcement learning-based trading strategies before deploying them in live trading environments.

# Chapter 5

# Expert

## 5.1 Explain the concept of stochastic volatility and how it is incorporated in option pricing models.

Stochastic volatility refers to the idea that the volatility of an asset price can vary randomly over time. This is in contrast to the traditional Black-Scholes model, which assumes that volatility is constant and known. Stochastic volatility models attempt to capture this randomness and uncertainty by introducing a stochastic process for volatility.

One popular model for stochastic volatility is the Heston model, which was introduced by Steven Heston in 1993. In the Heston model, the volatility of an asset is modeled as a mean-reverting process that is driven by a Brownian motion. Specifically, the model assumes that the volatility of the asset follows the following stochastic differential equation:

$$d\sigma_t = \kappa(\theta - \sigma_t)dt + \xi\sqrt{\sigma_t}dW_t^\sigma \tag{5.1}$$

where $\sigma_t$ is the volatility at time t, $\kappa$ is the rate of mean reversion,

$\theta$ is the long-term mean of the volatility, $\xi$ is the volatility of the volatility, and $W_t^\sigma$ is a Brownian motion that is correlated with the Brownian motion of the asset price.

When it comes to option pricing, stochastic volatility models can be used to price options by incorporating the randomness of volatility. One popular approach is to use Monte Carlo simulation, where the model is simulated many times to generate possible paths for the asset price and volatility. These paths are then used to calculate the expected value of the option payoff at expiration.

Another approach is to use partial differential equations (PDEs) to derive an analytical solution for the option price. This is typically done by applying the Feynman-Kac formula, which relates the option price to the solution of a PDE. The resulting PDE will depend on the particular stochastic volatility model being used, and may involve additional variables beyond the asset price and time.

Overall, stochastic volatility models provide a more realistic framework for option pricing by accounting for the randomness of volatility. While these models can be more complex than the traditional Black-Scholes model, they are widely used in quantitative trading and investment due to their ability to generate more accurate pricing and risk measures.

## 5.2   Describe the difference between local volatility models and stochastic volatility models in option pricing.

Local volatility models and stochastic volatility models are two approaches used to model the behavior of stock prices in option pricing. The main difference between the two types of models is the way they handle volatility.

Local volatility models assume that the volatility of the underlying asset is a deterministic function of time and price. In other words, the volatility is assumed to vary smoothly with the price and time, and its value at any given point in time is known. These models are also known as deterministic volatility models or Dupire models, named after Bruno Dupire who introduced the concept.

Stochastic volatility models, on the other hand, assume that the volatility of the underlying asset is a stochastic process. This means that the volatility is assumed to vary randomly over time and is not known with certainty at any given point in time. Stochastic volatility models are also known as random volatility models.

In local volatility models, the volatility surface is calibrated to market data to fit the observed prices of vanilla options. However, these models have limitations when it comes to pricing exotic options, such as barrier options or options with early exercise features. This is because these models assume that the volatility is known with certainty, which may not be the case in reality.

Stochastic volatility models, on the other hand, allow for more complex modeling of the volatility surface. There are several different types of stochastic volatility models, such as the Heston model, the SABR model, and the GARCH model. These models are more flexible in their assumptions about the volatility process and can be used to generate more accurate prices for exotic options.

In summary, both local volatility models and stochastic volatility models have their own strengths and weaknesses. Local volatility models are simpler to implement and can provide accurate prices for vanilla options, while stochastic volatility models are more flexible and can be used to price more complex exotic options.

## 5.3    What is the SABR model, and how is it used in interest rate derivative pricing?

The SABR model is a mathematical model used to price derivatives on assets with uncertain future volatility, such as interest rate derivatives. It was developed by Patrick Hagan, Deep Kumar, Andrew Lesniewski, and Diana Woodward in 2002 and is named after their initials.

The SABR model assumes that the underlying asset follows a lognormal stochastic process, with volatility following a stochastic process driven by a stochastic volatility factor. The model is characterized by four parameters: the initial value of the underlying asset, the volatility of the underlying asset, the correlation between the underlying asset and the stochastic volatility factor, and the volatility of the stochastic volatility factor.

The SABR model is particularly well-suited for pricing interest rate derivatives, which often exhibit a "smile" or "skew" in their implied volatility curves. This occurs because the volatility of interest rates is itself a function of the level of interest rates, with higher levels of interest rates generally implying greater volatility. The SABR model allows for this non-linearity in the relationship between interest rates and volatility by allowing the volatility of the stochastic volatility factor to vary with the level of interest rates.

The SABR model is commonly used in interest rate derivative pricing, particularly for options on interest rate swaps, caps, and floors. It has been widely adopted by banks and financial institutions for its flexibility and accuracy in capturing the volatility smile features of interest rate derivatives. The model is also utilized in hedging and risk management scenarios, where traders or fund managers can use the model to estimate the Greeks or sensitivities for their portfolios.

For example, a bank with an interest rate derivatives portfolio can use the SABR model to estimate its exposure to changes in interest rates and volatilities, and adjust its hedging strategy accordingly. Additionally, the model can help traders identify trading opportunities in the market by comparing current market prices to the model's implied prices.

## 5.4 Explain the Girsanov theorem and its application in the change of measure for risk-neutral valuation.

The Girsanov theorem is a fundamental result in probability theory that provides a method for changing the measure of a stochastic process. In the context of finance, the theorem can be used to derive the risk-neutral measure, which is a key concept in option pricing and quantitative trading.

The theorem states that if we have a stochastic process X(t) under the so-called "physical measure" P, then we can define a new measure Q that is equivalent to P, i.e., the probability of any event under P is the same as under Q, by specifying a exponential martingale as the change of measure, denoted by M(t):

$$M(t) = exp\left(B(t) - (\mu + 0.5 * \sigma^2)t\right) \tag{5.2}$$

where B(t) denotes a standard Brownian motion, and $\mu$ the drift and $\sigma$ volatility parameters are chosen such that the process X(t) becomes a Q-martingale, i.e., $E_Q[X(t)|F(s)] = X(s)$, for any $s \leq t$.

This means that the dynamics of X(t) under the risk-neutral measure Q are different from those under the physical measure P, but the expected values of X(t) are the same. By applying the Girsanov theorem, one can derive the risk-neutral drift and volatility of an asset's price process, which are used in option pricing and hedging.

For example, consider a European call option with strike price K expiring at time T on a stock with price process S(t). If we assume that the stock price follows a geometric Brownian motion under the physical measure, i.e., $dS(t) = \mu S(t)dt + \sigma S(t)dW(t)$, where $\mu$ is the drift rate, $\sigma$ is the volatility, and W(t) is a Brownian motion under P, then we can apply the Girsanov theorem to derive the corresponding stock price process under the risk-neutral measure Q:

$$dS(t) = rS(t)dt + \sigma S(t)dW_Q(t) \tag{5.3}$$

where r is the risk-free rate and $W_Q(t)$ is a Brownian motion under Q. The risk-neutral drift rate r replaces the physical measure drift rate , and is given by r = - R , where R is the market price of risk.

Using the risk-neutral price dynamics, one can then price the call option by taking the discounted expected value of the payoff under Q, and obtain the well-known Black-Scholes formula. This approach is known as risk-neutral valuation, and relies on the assumption that investors are risk-neutral and only care about the expected returns of their investments.

# 5.5 Describe the concept of no-arbitrage pricing and its importance in financial derivatives.

No-arbitrage pricing is a fundamental concept in financial derivatives that describes the idea that an asset's price should be equivalent to the sum of all future cash flows that the asset generates. The concept of no-arbitrage pricing is based on the idea that if two assets are equivalent in terms of their future cash flows, then their prices should be the same. If two assets have different prices, an arbitrage opportunity exists, and traders could make a risk-free profit by buying the lower-priced asset and selling the higher-priced asset.

No-arbitrage pricing is crucial in financial derivatives because it ensures that the price of a derivative is directly linked to the underlying asset's price. In simple terms, financial derivatives such as options, futures, and swaps are contracts that derive their value from the value of an underlying asset. Financial derivatives enable investors to speculate on an asset's future price without owning it, which means that the derivative's price must move in tandem with the underlying asset's price.

The concept of no-arbitrage pricing is important in financial derivatives because it ensures that derivatives' prices represent the expected future cash flows accurately. For instance, let us assume that a stock has a current price of $100 and a one-year forward price of $110. According to the concept of no-arbitrage pricing, an investor selling options on this stock must be compensated for the incremental risk he is taking on. Thus, the price of the option must be set at a level that reflects the risk of the underlying asset, plus an appropriate premium for the option seller.

No-arbitrage pricing is essential because it creates an efficient and fair financial system, where prices of assets are determined by their expected future cash flows. This system eliminates any possible arbitrage opportunities and ensures that investors are compensated fairly for the risks they assume in trading financial derivatives. In summary, No-arbitrage pricing is crucial in financial derivatives, as it ensures that the prices reflect the underlying assets' expected cash flows, creating an efficient and fair financial system.

## 5.6    What is the Heath-Jarrow-Morton (HJM) framework, and how is it used in interest rate modeling?

The Heath-Jarrow-Morton (HJM) framework is a widely-used mathematical model in finance that describes the evolution of interest rates over time. It was introduced in a seminal paper by David Heath, Robert Jarrow, and Andrew Morton in 1992.

In its simplest form, the HJM framework is a stochastic differential equation that models the dynamics of the entire yield curve, from short-term rates to long-term rates. The model assumes that interest rates are continuous and follow a lognormal diffusion process, and that the shape of the yield curve is stationary over time.

The HJM framework is used extensively by financial institutions and investors for interest rate modeling, risk management, and trading. Some applications of the model include:

1. Interest rate forecasting: The HJM model can be used to forecast the future movements of interest rates, providing valuable information to traders and investors.

2. Portfolio optimization: The HJM model can be used to optimize portfolios of bonds or other interest-rate sensitive securities. By forecasting interest rates, investors can adjust their holdings to maximize returns while minimizing risk.

3. Derivative pricing: The HJM model is often used to price interest rate derivatives, such as swaps and swaptions. The model can be used to simulate interest rate paths under different scenarios, allowing traders to price and hedge complex derivative instruments.

4. Risk management: The HJM model is used extensively in risk management to measure and mitigate interest rate risk. By forecasting future interest rates, investors can adjust their portfolios to protect against adverse interest rate movements.

Overall, the HJM framework is a powerful tool for understanding and modeling interest rate dynamics. By providing a comprehensive and flexible model of the yield curve, the HJM framework has become a staple in financial markets and has enabled new insights and strategies for traders and investors.

# 5.7 Explain the concept of a forward measure and its application in fixed income securities.

In probability theory and quantitative finance, a forward measure is a measure that is used to discount future cash flows. It provides a way to value financial assets and derivatives that depend on uncertain future events, such as fixed income securities like bonds.

The main advantage of using a forward measure is that it takes into account the time value of money by assuming a risk-neutral investor who does not require a risk premium for holding an asset. This measure is constructed by assuming that the expected rate of return on an investment is equal to the risk-free rate, which is the rate of return on a risk-free asset such as a government bond.

In fixed income securities, the forward measure is often used for pricing and valuing interest rate swaps, fixed-rate bonds, and other derivative instruments. The forward rate agreement (FRA) is a common application of the forward measure, which is used to hedge against interest rate risk.

For example, let's say an investor purchases a fixed-rate bond that pays a coupon of 5% p.a. over five years. The bond price would be calculated based on the forward measure that considers the expected interest rates over the next five years. The interest rates are predicted using the yield curve, which is a graph that shows the interest rates for different maturities.

If the forward measure indicates that the interest rates are expected to increase over the next five years, the bond price will be lower than the face value, reflecting the higher interest rates. On the other hand, if the forward measure indicates that the interest rates are expected to decrease, the bond price will be higher than the face value.

In summary, the forward measure is a useful concept in valuation and risk management of fixed income securities. It enables market participants to price assets and derivatives that depend on future events and to manage the associated interest rate risk.

## 5.8 Describe the Vasicek model and its application in interest rate modeling.

The Vasicek model is a mathematical model used to describe the behavior of interest rates over time. It was developed by Oldrich Vasicek in 1977 and is one of the earliest and most influential models of interest rate dynamics. The model assumes that the short-term interest rate follows a stochastic process that is mean-reverting, where the rate tends to revert to its long-term average over time.

Mathematically, the Vasicek model is described by the following differential equation:

$$dr(t) = (a - b * r(t))dt + \sigma * dW(t) \tag{5.4}$$

Where:

```
- r(t) is the short-term interest rate at time t
- a is the long-term average interest rate
- b is the speed of reversion to the mean
- is the volatility of the interest rate
- dW(t) is a Wiener process, which represents the random fluctuations in the
      interest rate over time.
```

The model assumes that interest rates are normally distributed, which allows for the calculation of probabilities and the pricing of interest rate options using the principles of probability theory.

The Vasicek model has been widely used in interest rate modeling and is particularly useful in pricing interest rate derivatives, such as interest rate swaps and caps/floors. For example, the model can be used to calculate the probability of interest rates reaching a certain level or to estimate the value of an interest rate option based on the expected future behavior of interest rates.

One limitation of the Vasicek model is that it assumes interest rates are capable of negative values, which in reality is not possible. This has led to the development of other interest rate models, such as the Cox-Ingersoll-Ross (CIR) model and the Hull-White model, which address this limitation.

## 5.9    Explain the concept of the Hull-White model and its use in interest rate derivative pricing.

The Hull-White model is a mathematical model used in finance to describe interest rate movements over time. It was developed by John Hull and Alan White in the 1990s as an extension of the Vasicek model.

The basic idea behind the Hull-White model is that interest rates are driven by both a long-term equilibrium level and short-term fluctuations. The model assumes that short-term interest rates follow a mean-reverting process, which means that they tend to return over time to their long-term equilibrium level.

The Hull-White model is typically used in interest rate derivative pricing because it provides an efficient way to value complex financial instruments like options and swaptions. By modeling interest rate movements over time, the Hull-White model can generate simulated paths for interest rates, which can then be used to value these derivative instruments.

One advantage of the Hull-White model is that it allows for the modeling of both deterministic and stochastic interest rate movements. This is important because real-world interest rates can vary for a variety of reasons, including market trends, economic fundamentals, and policy changes.

Another advantage of the Hull-White model is its flexibility. The model can be calibrated to match a wide range of interest rate data, including yield curves and volatility surfaces. This makes it a useful tool for analyzing and valuing interest rate derivatives in a variety of market conditions.

Overall, the Hull-White model is an important tool for interest rate derivative pricing and risk management. By providing a mathematical framework for modeling interest rate movements, it helps traders and investors make informed decisions in a complex and rapidly changing market.

# 5.10 Describe the concept of the LIBOR Market Model (LMM) and its application in interest rate derivative pricing.

The LIBOR Market Model (LMM) is a mathematical model used to price interest rate derivatives, which are financial instruments whose value is derived from the fluctuations in interest rates. The LMM was developed to model the evolution of the term structure of interest rates using an approach based on the evolution of the underlying short-term interest rates, which are represented by the London Interbank Offered Rate (LIBOR).

The LMM is a forward-looking model that takes into account the dynamics of interest rates over time, including the volatility of the rates and the correlation between different rates. The model is typically used to price complex interest rate derivatives, such as caps, floors, swaptions, and other exotic instruments.

One of the key benefits of using the LMM is that it allows for the modeling of interest rate volatility and correlation. This is important because it enables market participants to better understand and manage the risks associated with interest rate derivatives. The volatility and correlation parameters can be estimated from market data and used to simulate future interest rate scenarios, which can be used to price derivatives and hedge portfolios.

The LMM can also incorporate the impact of macroeconomic variables on the term structure of interest rates. For example, changes in inflation or economic growth can affect interest rates in different ways, and the LMM can capture these effects in its simulations.

In practice, the LMM is often combined with Monte Carlo simulation techniques to generate a large number of possible interest rate scenarios. These scenarios can then be used to price complex interest rate derivatives and to estimate the associated risks.

Overall, the LMM is a powerful tool for pricing and managing interest rate derivatives. Its ability to capture the complexity of interest rate dynamics and to incorporate macroeconomic variables make it an important tool for market participants in today's financial markets.

## 5.11    What is the difference between risk-neutral and real-world probability measures in quantitative finance?

In quantitative finance, the two most common types of probability measures used are risk-neutral and real-world probabilities.

Risk-neutral probability measures are used when pricing derivatives, such as options, in the financial markets. These probabilities represent the probability of different future outcomes under an assumption that investors are indifferent to risk. In other words, in the risk-neutral world, investors only care about the expected return of an investment, and not the potential risks associated with it.

Real-world probability measures, on the other hand, represent the actual probabilities of different future outcomes. These probabilities take into account the risks associated with an investment and reflect the actual likelihood of those risks materializing.

The difference between risk-neutral and real-world probabilities is important because it influences the pricing of derivatives. When using risk-neutral probabilities, the price of a derivative is equal to its expected payout discounted at the risk-free rate. Real-world probabilities, however, may differ from risk-neutral probabilities due to risks associated with an asset, such as credit risk, liquidity risk or market risk. These risks will affect the pricing of the asset in the real world and may lead to differences between the prices obtained using real-world and risk-neutral probabilities.

To illustrate, consider the pricing of a European call option on a stock. Under the risk-neutral measure, the price of the call option is equal to the discounted expected value of the option payoff at the option expiry date. In contrast, under the real-world measure, the price of the call option will depend on the volatility of the stock price and whether the option is in or out of the money.

In conclusion, the difference between risk-neutral and real-world probabilities is important in quantitative finance, as it affects the pricing of financial instruments, particularly derivatives. Risk-neutral probabilities are used to price derivatives, while real-world probabilities reflect the actual risks and may affect the pricing of the assets in the real world.

## 5.12 Explain the use of Kalman filtering in state space models for time series analysis.

Kalman filtering is a mathematical filtering and smoothing technique that is widely used in state space modeling of time series data. State space models are mathematical models used to describe the evolution of a system over time in terms of repeated measurements or observations. They are used to model a wide range of time series situations, including financial and economic data.

In a state space model, the underlying state of the system is unobserved, and only a noisy and imperfect measurement of the state is available. The Kalman filter is an algorithm that uses this observed data to make estimates of the unobserved state of the system over time. It utilizes a recursive algorithm that incorporates new observations as they become available to provide updated estimates of the state at each time point.

In finance, Kalman filtering is used to filter out noisy market data and extract informative signals from the observations. This information can then be used to develop trading strategies and make investment decisions. For example, a Kalman filter can be applied to estimate the trend and volatility of a stock price series. The estimated trend can be used to construct a momentum trading strategy while the estimated volatility can be used to construct a volatility-based trading strategy.

Kalman filtering is also used extensively in quantitative finance to build models for asset pricing, portfolio optimization, risk management, and asset allocation. Kalman filters are typically used in combination with other quantitative techniques such as regression analysis, time series analysis, and machine learning, to develop robust and predictive models of financial time series data.

Overall, Kalman filtering is a powerful tool for state space modeling
and time series analysis in finance and can help traders and investors
to make better decisions, reduce risk, and increase returns.

## 5.13    How do you apply advanced optimization techniques, such as genetic algorithms and simulated annealing, in portfolio optimization?

Advanced optimization techniques, such as genetic algorithms and
simulated annealing, can be applied in portfolio optimization to search
for the optimal portfolio that maximizes returns or minimizes risk.
They allow the investor or trader to find a solution that is potentially
better than a simple heuristic solution, and provide a useful starting
point for further refinement.

Genetic algorithms are a type of optimization algorithm that mimic
the process of natural selection. It involves the use of a population
of potential outcomes (portfolios), where each potential outcome is
represented as a chromosome. Each chromosome contains a set of decision variables which represent the weightings of different securities
in the portfolio. The genetic algorithm then applies a set of evolu-
tionary operators, including crossover, mutation, and selection, to the
population to produce a new set of potential outcomes. The process
is then repeated until the algorithm converges to a solution.

Simulated annealing, on the other hand, is a stochastic optimization
technique that is inspired by the cooling process in metallurgy. Simulated annealing involves starting with a starting solution and repeatedly perturbing the solution by making small random changes (e.g.
adjusting portfolio weightings). The algorithm then accepts the new
solution if it improves the objective function (e.g. increases portfolio
returns or lowers portfolio risk), but also accepts worse solutions with
a probability that decreases over time. The algorithm continues this
process until it converges to a local or global optimum.

The basic idea behind portfolio optimization is to select a set of assets that will yield the highest returns for a given level of risk. One

common method of portfolio optimization that can be applied using advanced optimization techniques is Markowitz's mean-variance optimization. This method involves minimizing the portfolio variance subject to a specified expected return. In this case, the expected return is the mean return of the portfolio, and portfolio variance is the measure of portfolio risk.

Both genetic algorithms and simulated annealing can be used to solve this optimization problem. The algorithms can generate a set of potential portfolios that meet the investor's constraints (e.g. risk tolerance, investment horizon, etc.). The investor can then choose the portfolio that best fits their investment objectives.

For instance, assume a portfolio is high-risk as it consists of stock options but has a high return. An investor decides to optimize the portfolio in a way that decreases the risk but also maintains the high returns. With a genetic algorithm, different portfolios are created from randomization and mutated repeatedly until the best solution with optimal returns is obtained. If simulated annealing is used, the solution becomes optimized by decreasing the risk and can also maintain the high returns of the portfolio.

## 5.14 Describe the concept of market microstructure and its relevance in high-frequency trading.

Market microstructure refers to the study of the process and mechanisms by which assets are traded in financial markets. It encompasses the rules and protocols governing markets, the market participants, their interactions and their impact on prices.

In high-frequency trading, market microstructure is highly relevant due to the nature of this trading activity which involves using complex algorithms and powerful computing tools to analyze and execute trades at a very high frequency and speed. High-frequency traders rely on market microstructure to gain an advantage over other market participants by exploiting small price discrepancies and temporary imbalances in supply and demand.

Some of the key concepts in market microstructure that are highly relevant to high-frequency trading include:

- Order book: The order book is the electronic record of all pending buy and sell orders for a particular asset. High-frequency traders often analyze the order book to identify trends and patterns in the flow of orders and to adjust their trading strategies accordingly.

- Bid-ask spread: The bid-ask spread is the difference between the highest price a buyer is willing to pay for an asset (the bid price) and the lowest price a seller is willing to accept (the ask price). High-frequency traders often seek to profit by exploiting small fluctuations in bid-ask spreads.

- Market liquidity: Liquidity refers to the ease with which an asset can be bought or sold without affecting its market price. High-frequency traders often focus on highly liquid assets that can be traded quickly and easily without causing significant price movements.

- Market impact: The market impact is the effect of a large trade on the market price of an asset. High-frequency traders often aim to minimize their market impact by executing trades in small sizes over short time periods.

Overall, market microstructure is a crucial element of high-frequency trading, as it provides a detailed understanding of the structure and dynamics of financial markets, and can be used to develop and optimize trading strategies.

## 5.15 Explain the role of limit order books and order flow in market microstructure research.

Limit order books and order flow are two important concepts in market microstructure research, which studies the behavior of financial markets and how they operate at a very granular level.

A limit order book is a record of all the outstanding buy and sell orders in a market, along with the prices at which they are willing to trade. It contains limit orders, which are buy or sell orders with specific prices and quantities. The limit order book is a fundamental component of market microstructure research, as it reflects the supply and demand dynamics of the market, and contains valuable information about the preferences and strategies of traders.

Order flow, on the other hand, refers to the sequence of buy and sell orders that are placed in the market. Order flow is the major driver of changes in the limit order book, and can reveal important insights about the liquidity, volatility, and efficiency of a market. Understanding order flow is critical for designing effective trading strategies, as traders can use this information to anticipate market movements and adjust their positions accordingly.

In market microstructure research, analysts use sophisticated statistical models and algorithms to analyze and interpret limit order book data and order flow patterns. They use this information to develop trading strategies that take advantage of expected price movements, and to measure the impact of different market conditions on trading performance. For example, by analyzing the limit order book, researchers can identify patterns of price clustering, which indicate that traders are using certain price levels as support or resistance. These insights can inform the design of trading algorithms that aim to exploit these patterns for profit.

Overall, the study of market microstructure and the use of limit order book and order flow data are critical for understanding the complexities of financial markets and designing effective trading strategies.

## 5.16   What is the concept of optimal execution in algorithmic trading, and what are some common strategies?

Optimal execution is the process of buying or selling a security while minimizing the market impact and transaction costs associated with the trade. In algorithmic trading, optimal execution is a crucial component of any trading strategy as it helps traders achieve the best possible price for a given trade.

There are several common strategies used in optimal execution, including:

1. VWAP (Volume-Weighted Average Price): This strategy aims to buy

or sell a security at a price that is close to the average price weighted by trade volumes over a specific time frame. This strategy is useful when trading in large sizes with low market impact.

2. TWAP (Time-Weighted Average Price): This strategy aims to buy or sell in equal quantities over time, which can minimize both market impact and transaction costs. This strategy is useful when executing small to medium-sized trades.

3. Implementation Shortfall: This strategy measures the difference between the initial price and the final execution price, taking into account market movements and trading costs. This strategy aims to minimize losses and maximize profits while executing trades.

4. Sniper: This strategy focuses on waiting for market inefficiencies or mispricings, then making trades to exploit those opportunities quickly. This strategy requires fast reaction times and relies heavily on algorithms and automation.

5. Liquidity-seeking: This strategy is designed to minimize market impact by actively seeking out the most liquid areas of the market to execute trades. This strategy can be useful when executing larger orders.

Overall, the key to optimal execution in algorithmic trading is to carefully balance the need for quick execution with the need to minimize market impact and trading costs. A well-designed trading algorithm can help traders achieve these goals and improve overall performance.

# 5.17   Describe the use of text analysis and natural language processing (NLP) in quantitative finance.

Text analysis and natural language processing (NLP) are powerful tools that can be used in quantitative finance to extract meaningful insights about financial markets from unstructured text data. Here are some specific examples of how these techniques are used:

1. Sentiment Analysis: Sentiment analysis is the process of determining the emotional tone behind a piece of text. In finance, sentiment analysis can be used to gauge investors' emotions and expectations surrounding a particular stock, commodity or currency. A positive sentiment score may indicate an upward trend or positive outlook, while a negative score may indicate a downward trend or negative outlook. This can be used by

quantitative traders as a factor in deciding when to buy or sell an asset.

2. News Events Detection: NLP can aid in detecting significant news events related to a stock or industry that can potentially impact its value. For instance, when a company announces significant progress on a drug trial, this can positively affect its stock price. Text analysis can help to quickly identify and analyze news articles to extract important financial information, enabling traders to make quick and informed trading decisions.

3. Text-based Factor Analysis: Text-based factor analysis is a technique that combines quantitative data analysis with information extraction from natural language sources. This can be useful in predicting rare events that are not captured in traditional financial models. For instance, if a key executive leaves a company, this may not show up in a company's financial statements, but through text analysis, this information can be gathered to indicate potential future underperformance or changes in the company's valuation.

4. Macro-level Analysis: NLP techniques can be used to sift through vast amounts of text data to identify global trends and macroeconomic events that may impact financial markets around the world. This can help traders make informed decisions that take into account the broader global context.

Overall, the use of text analysis and natural language processing in quantitative finance is an innovative approach that is changing how investment decisions are made. By harnessing the power of language, traders are able to capture important information that might be missed using traditional quantitative methods.

## 5.18    Explain the concept of regime-switching models and their application in financial markets.

Regime-switching models are a type of statistical model used to describe the behavior of time-series data that changes its characteristics over time in response to different market conditions or regime shifts. These models are useful in financial markets because the relationship between different asset classes and their volatility is often highly dependent on the current economic environment, which can change rapidly and unpredictably.

In a regime-switching model, the underlying time-series data is assumed to be generated by a series of different stochastic processes, each of which corresponds to a different market regime. These regimes can be thought of as distinct states that are characterized by different levels of mean return, volatility, and correlation among asset classes.

The key concept behind regime-switching models is the idea of a latent variable or state that determines the current market regime at any given time. The state is typically assumed to follow a Markov process, which means that the probability of transitioning from one state to another depends only on the current state and not on any previous states.

One common example of a regime-switching model is the Hidden Markov Model (HMM), which assumes that the observed data is generated by an unobserved Markov process that determines the current regime. HMMs have been used extensively in finance to model asset prices and volatility, as well as to identify different market regimes such as bull and bear markets, high and low volatility periods, and different economic cycles.

In financial markets, regime-switching models can be used for a variety of applications including risk management, portfolio optimization, and trading strategies. For instance, an investor might use a regime-switching model to adjust their portfolio allocation depending on the current market regime. In a bear market, the model might suggest a heavier allocation to defensive assets like bonds, cash, or gold, while in a bull market it might suggest a more aggressive allocation to equities or other risk assets.

Overall, regime-switching models provide a flexible and powerful framework for modeling the complex and dynamic behavior of financial markets, which can help investors better understand and manage risk. By taking into account the possibility of regime shifts and the different market states that exist, these models can offer deeper insights into the underlying dynamics of financial markets and help investors make more informed investment decisions.

## 5.19 How do you incorporate transaction costs and other frictions in portfolio optimization and trading strategies?

Transaction costs and other frictions are vital considerations when designing portfolio optimization and trading strategies. These costs can eat up a significant portion of potential profits and can easily turn a profitable strategy into a losing one.

There are several ways to incorporate transaction costs and other frictions into portfolio optimization and trading strategies. Below are some of the most common methods:

1. Use simulation: Simulation is the most common approach to dealing with transaction costs and other frictions in portfolio optimization and trading strategies. This approach involves running simulations that incorporate all relevant costs and frictions. By doing this, you can estimate the impact of transaction costs and other frictions on your portfolio returns, and use this information to adjust your trading strategies.

2. Consider slippage: Slippage is the difference between the price at which you intended to execute a trade and the actual price at which it was executed. This can be significant in fast-moving markets or when trying to trade large positions. One way to accommodate slippage is to incorporate it into the cost of the trade. For example, you could assume that you buy at the mid-price plus a small premium and sell at the mid-price minus a small discount.

3. Use limit orders: Limit orders can be helpful in reducing transaction costs and slippage. A limit order is an order to buy or sell a security at a specific price or better. By using limit orders, you can potentially avoid high transaction costs associated with market orders or reduce slippage by ensuring that trades are executed at a predetermined price.

4. Consider market impact: The market impact refers to the impact of your trades on the market. Trading large positions or illiquid securities can move the market and cause the price to move against you. When incorporating market impact into your portfolio optimization and trading strategies you should consider the impact on transaction costs, slippage and overall portfolio performance.

5. Use optimization algorithms: Modern portfolio optimization techniques incorporate transaction costs and other frictions such as taxes, fees, and bid-ask spreads into their calculations. These optimization algorithms take into account the impact of transaction costs and other frictions, and

deliver optimized portfolios that account for these costs.

For example, consider a portfolio manager who charges a management fee of 1% per year and incurs an average transaction cost of 0.25% for each trade. If the portfolio manager makes five trades per year, the total transaction costs would be 1.25%. To account for this, the portfolio manager must earn an additional 1.25% just to break even. By incorporating these costs into the optimization process, the portfolio manager can create a portfolio that accounts for these costs and delivers an appropriate return to the investor.

In conclusion, transaction costs and other frictions should be taken into account when designing portfolio optimization and trading strategies. Failing to account for these costs can mean the difference between a successful strategy and a losing one. It is important to use simulation, consider slippage, use limit orders, consider market impact and use optimization algorithms to incorporate these costs into your strategies.

## 5.20  What are some common methods for evaluating the performance of quantitative trading strategies, and how do you account for overfitting?

When evaluating the performance of quantitative trading strategies, it is essential to consider various metrics to assess their effectiveness. Some of the most common techniques used for this evaluation include backtesting, live trading, simulation, and stress testing.

Backtesting is the process of evaluating a trading strategy using historical financial data. This method involves running the strategy on past data to determine how it would have performed and identify potential issues in the strategy. Additionally, backtesting enables traders to refine their strategies by adjusting various parameters on the basis of past performance.

Live trading involves implementing a strategy in real-time financial markets. This method provides traders with the opportunity to validate their strategy's accuracy and effectiveness in a real-world environment. It is crucial to monitor the live trading strategies continuously and adjust them if necessary.

Simulation involves creating a computer model to simulate financial market conditions and the effects of trading strategies. This method allows traders to assess the performance of a strategy over a long period and make changes to the strategy before risking real capital.

Stress testing involves subjecting the strategy to worst-case scenario conditions, such as sudden changes in the financial market or a significant increase in volatility. Stress testing enables traders to determine the resilience of their strategy under adverse conditions.

Overfitting is a common issue in quantitative trading strategies, where the strategy performs well on historical data but fails in the real market. To avoid overfitting, it is essential to validate the strategy using out-of-sample data, which involves testing the strategy on financial data that were not used during the backtesting process. Additionally, traders can use techniques such as cross-validation, regularization, and ensemble methods to reduce the risk of overfitting.

In summary, traders must use a combination of backtesting, live trading, simulation, and stress testing to evaluate the performance of their quantitative trading strategies accurately. They must also account for overfitting by validating their strategy using out-of-sample data and implementing techniques to reduce the risk of overfitting.

# Chapter 6

# Guru

## 6.1 Describe the concept of rough volatility models and their application in option pricing.

Rough volatility models are a class of stochastic volatility models that are based on fractional Brownian motion. In contrast to traditional stochastic volatility models, which assume smooth and continuous paths for the volatility process, rough volatility models assume that the volatility process is rough and exhibits long-range dependence.

The roughness of the volatility process in rough volatility models is controlled by the Hurst parameter, which is related to the degree of persistence or memory in the volatility process. A Hurst parameter value of 0.5 corresponds to Brownian motion, while values greater than 0.5 indicate long-range dependence and increased volatility clustering.

One of the key applications of rough volatility models is in option pricing. Traditional stochastic volatility models assume that the volatility of the underlying asset is constant and can be modeled by a single stochastic process. However, in reality, volatility is not constant and exhibits significant fluctuations over time. Rough volatility models provide a more accurate description of the volatility process and allow for more realistic option pricing.

Rough volatility models can be used to derive option pricing formulas that are more accurate and computationally efficient than traditional models. For example, in the Heston model, which is a popular stochastic volatility model, the volatility process is assumed to follow a specific stochastic differential equation (SDE). In contrast, rough volatility models allow for more flexible and realistic volatility dynamics that are not restricted to a specific SDE.

In addition, rough volatility models can also be used for risk management and hedging strategies. By accurately modeling the volatility process, traders and investors can better manage their risk exposure and develop more effective hedging strategies.

Overall, rough volatility models represent an important development in option pricing and quantitative finance more broadly, as they provide a more accurate and realistic description of financial market dynamics.

## 6.2   Explain the use of model-free implied volatility measures, such as the VIX index, in quantitative finance.

Model-free implied volatility measures, such as the CBOE Volatility Index (VIX), are used extensively in quantitative finance for a variety of purposes. The VIX is a measure of the market's expectation of 30-day implied volatility for the S&P 500 index options, and it is calculated using a formula that is based on the prices of out-of-the-money (OTM) put and call options on the S&P 500.

One of the primary uses of the VIX is as a measure of market risk. When the VIX is high, it suggests that market participants are expecting increased volatility and uncertainty in the future. This can be a useful signal for traders and investors to adjust their positions and risk-management strategies accordingly. For example, a high VIX might prompt an investor to reduce their exposure to equities and increase their allocation to bonds or cash.

Another use of the VIX is as a trading instrument in its own right. The popularity of exchange-traded products (ETPs) that are based on the VIX, such as VIX futures and options, has grown significantly in recent years. These products allow traders and investors to take positions on the future level of volatility in the market, without having to trade options on individual stocks or indices.

The VIX and other model-free implied volatility measures can also be used to develop trading strategies that exploit market mispricings. For example, if the VIX is higher than the implied volatility of individual stocks in the S&P 500, a trader might bet on the VIX falling and individual stock volatilities rising. This strategy would involve buying individual stock options while shorting VIX futures or options.

It is worth noting that while model-free implied volatility measures such as the VIX are useful tools for understanding market risk and developing trading strategies, they are not without limitations. For example, they are based on assumptions about market participants' expectations of future volatility that may not always be accurate. Therefore, it is important to use these measures in conjunction with other analysis and risk-management techniques when making investment decisions.

## 6.3   How do you incorporate market frictions, such as liquidity and transaction costs, in derivative pricing models?

When pricing derivatives, it is important to incorporate market frictions such as liquidity and transaction costs in order to accurately assess the market price of the underlying asset. Market frictions affect the supply and demand of the asset, as well as the costs associated with trading. Derivatives pricing models typically incorporate these frictions through the use of adjusted pricing formulas or simulations.

One common approach is to adjust the cost of carry in the pricing model to account for transaction costs and other market frictions. In the Black-Scholes model, for example, the cost of carry is the risk-free interest rate minus the dividend yield. In the presence of market frictions, this formula is adjusted to include the costs associated with trading the underlying asset. This results in a modified Black-Scholes model that takes into account transaction costs and other expenses.

Another approach is to incorporate market frictions into simulations, such as Monte Carlo simulations. These simulations can include variables such as bid-ask spreads and order book depth, which affect the liquidity and transaction costs of trading the underlying asset. By running simulations with different sets of market frictions, traders can evaluate the impact of these frictions on the pricing of the derivative.

To illustrate the impact of market frictions on derivative pricing, lets consider an example of a call option on a stock. In the absence of market frictions, the Black-Scholes model can be used to price the option. However, if the stock is illiquid and has a large bid-ask spread, the pricing model must be modified to account for these frictions. This can result in a lower valuation for the option due to the added costs of executing the trade.

In conclusion, market frictions such as liquidity and transaction costs can significantly impact the pricing of derivatives. Pricing models must be adjusted to incorporate these costs in order to accurately assess the market price of the underlying asset. Simulation-based approaches can also be used to evaluate the impact of these frictions on the derivative pricing. Incorporating market frictions into derivative pricing models is crucial for effective risk management and trading strategies.

## 6.4   Describe the concept of a change of numéraire and its application in fixed income and derivative pricing.

Change of numeraire is a concept used in quantitative finance for pricing derivatives or fixed income securities. It involves changing the underlying asset used to value the instrument.

In traditional derivative pricing, the pricing is done using a single underlying asset, but in reality, the value of the asset can have high variability that may make the pricing process difficult. Therefore, changing the underlying asset can simplify the pricing process by reducing volatility or making it more manageable.

For example, suppose we have a European call option that expires in one year, with a strike price of $100, and the underlying asset is a stock. Let's assume that the volatility of the stock is very high, and the market is not very stable. In this case, we can choose to change the underlying asset from the stock to a fixed income instrument, such as a zero-coupon bond. We could use the bond as the numeraire and value the option based on its payout relative to the bond.

By choosing a less volatile and more predictable asset as the pricing reference, we can produce a more stable valuation of the derivative. We can show how the option is valued relative to the bond, allowing for more transparent pricing through the expected values of the payoff due at different times.

Similarly, the concept of a change of numeraire can also be applied to fixed income securities. Fixed-income securities are typically priced using a default-free bond as the benchmark. In this context, the numeraire is the interest rate. However, when interest rates become volatile or behave in unexpected ways, this standard framework can also become problematic. To address this, alternative numerairess are used for the valuation of fixed-income securities. These numerairess may include inflation rates, commodity prices, or foreign exchange rates, among others.

Overall, changing the numeraire can help reduce complex pricing equations and make the valuation of derivatives or fixed income securities more manageable. It also leads to more consistent pricing by using a less volatile and more predictable asset.

## 6.5 Explain the martingale representation theorem and its significance in quantitative finance.

The martingale representation theorem is a fundamental result in probability theory and has significant implications for quantitative finance. The theorem states that any martingale process (i.e., a stochastic process that satisfies certain mathematical conditions) can be represented as the sum of a predictable process (a process whose future values can be predicted with certainty based on its past values) and a martingale starting from zero.

More formally, let $(F, P)$ be a probability space and let $X_t$ be a martingale with respect to a filtration $F_t$. Then, there exists a predictable process $H_t$ such that $X_t = H_t + M_t$, where $M_t$ is a martingale starting from zero.

In the context of quantitative finance, the martingale representation theorem has important implications for pricing and hedging financial instruments. The theorem implies that any financial instrument whose price can be modeled as a martingale process can be expressed as a sum of a predictable component and a martingale component. The predictable component can be thought of as the deterministic part of the price, and the martingale component can be thought of as the stochastic part.

For example, consider a European call option on a stock whose price is modeled as a geometric Brownian motion. The price of the option can be expressed as a martingale process, which means that it satisfies the conditions of the martingale representation theorem. Using the theorem, we can express the price of the option as the sum of a predictable component (the present value of the expected payoff) and a martingale component (the stochastic variation in the stock price).

The martingale representation theorem is also related to the concept of risk-neutral pricing. In a risk-neutral world, the expected return on any financial instrument is equal to the risk-free rate. The martingale representation theorem provides a mathematical framework for constructing risk-neutral pricing models.

In summary, the martingale representation theorem is a powerful tool in quantitative finance that allows us to express the price of financial instruments in terms of a predictable component and a stochastic component. This has important implications for pricing and hedging financial instruments, as well as for constructing risk-neutral pricing models.

## 6.6 What is the concept of affine term structure models, and how are they used in interest rate modeling?

Affine term structure models (ATSMs) are a popular class of models used to describe the term structure of interest rates. These models are based on the assumption that the logarithm of the yield curve can be represented as a linear combination of a constant, a set of functions of time, and a set of normally distributed random variables.

The ATSM allows for the modeling of both the mean and volatility of the yield curve by specifying the time-varying parameters of the model. The model parameters are typically estimated using historical yield curve data, and can be used to price fixed income securities, as well as to value interest rate derivatives.

One of the key advantages of ATSMs is their flexibility in accommodating a wide range of interest rate dynamics. For example, the popular Vasicek and Cox-Ingersoll-Ross (CIR) models are both special cases of ATSMs, and can be used to model interest rate processes with mean reversion and volatility clustering.

ATSMs are commonly used in quantitative finance for a variety of applications such as modeling interest rate term-structures, pricing fixed income securities, developing fixed income trading strategies, and risk management. They are also useful in the analysis of market expectations of future interest rates, as well as in pricing and managing credit derivatives.

For example, an investment bank that engages in fixed income trading might use an ATSM for pricing complex interest rate derivatives such as swaps, options, and swaptions. This could involve simulating future interest rate scenarios to assess the risk and profitability of various trading strategies. Similarly, a pension fund might use an ATSM to model the behavior of interest rates in order to better manage the interest rate risk associated with their fixed income investments.

In summary, Affine term structure models are widely used in interest rate modeling and offer a flexible and powerful framework for analyzing fixed income markets. They are particularly useful for pricing and hedging interest rate derivatives and for conducting risk management in fixed income portfolios.

## 6.7   Describe the use of neural network-based calibration methods for complex derivative pricing models.

Neural network-based calibration methods have become increasingly popular in recent years for calibrating complex derivative pricing models. These methods involve the use of machine learning techniques to learn the mapping between the model input parameters and the model output, which is typically the price of the derivative.

To use neural network-based calibration methods, one must first define a training data set, which is typically generated using some numerical method, such as Monte Carlo simulation. This data set should include a range of input parameter values and their corresponding outputs (i.e., prices) from the pricing model.

Once the training data set is defined, a neural network architecture can be created and trained using the data set. The architecture typically consists of an input layer, one or more hidden layers, and an output layer, with the number of nodes in each layer determined by the complexity of the pricing model and the size of the training data set. The network is trained using a backpropagation algorithm to minimize the difference between the predicted prices and the actual prices from the data set.

After the training is complete, the neural network can be used to calibrate the pricing model by feeding in new input parameter values and obtaining the corresponding price prediction from the network. This provides an efficient and accurate way to calibrate complex pricing models, particularly when traditional numerical methods are computationally expensive or unstable.

One example of a neural network-based calibration method is the Longstaff-Schwartz neural network regression method, which uses a feedforward neural network to predict the continuation value of an American option, and then uses that prediction to estimate the option price. Another example is the Deep Hedging method, which uses deep neural networks to optimize hedging strategies for complex derivative portfolios.

Overall, neural network-based calibration methods offer a promising approach to pricing complex derivatives, particularly in cases where traditional numerical methods may be prohibitively expensive or unreliable.

## 6.8   Explain the concept of utility maximization and its role in optimal portfolio selection.

Utility maximization is a concept in finance that suggests investors make decisions based on how much "utility" they will receive from a particular investment. This utility is usually related to the level of satisfaction or happiness that the investor feels as a result of the return on the investment.

The concept of utility maximization is important in optimal portfolio selection because investors want to balance the trade-off between risk and return in their portfolio. They aim to find the optimal combination of assets that will provide the highest level of utility for a given level of risk.

The process of portfolio selection involves assessing the risk and return characteristics of various assets and constructing a portfolio that maximizes the investor's utility. To do this, investors use a variety of tools, including mean-variance analysis, optimization models, and simulation techniques.

Mean-variance analysis is a popular method used to construct optimal portfolios. The analysis involves plotting the expected return and risk for each asset in the portfolio and identifying the set of portfolios that provide the highest level of return for a given level of risk. This set of portfolios is known as the efficient frontier.

Once the efficient frontier has been identified, investors can use optimization models to select the portfolio that provides the highest level of utility for their risk preferences. Optimization models employ mathematical algorithms to find the portfolio that maximizes the investor's utility function.

Furthermore, simulation techniques can also be used to estimate the probability of different outcomes for a given set of asset returns. By simulating different scenarios, investors can test the robustness of their portfolio selection and identify potential downside risks.

In summary, utility maximization plays a crucial role in optimal portfolio selection by helping investors balance the trade-off between risk and return. The goal is to construct a portfolio that provides the highest level of utility for a given level of risk by using various analytical tools, including mean-variance analysis, optimization models, and simulation techniques.

## 6.9    How do you incorporate higher moments, such as skewness and kurtosis, in portfolio optimization and risk management?

Incorporating higher moments, such as skewness and kurtosis, in portfolio optimization and risk management provides investors with more information about the distribution of returns and allows for improved risk management. Higher moments can be useful in understanding the non-normality of returns and in constructing portfolios that are more robust to extreme events.

Here are a few ways to incorporate higher moments in portfolio optimization and risk management:

1. Use a downside risk measure: Downside risk measures, such as Value-at-Risk (VaR) and Expected Shortfall (ES), take into account the skewness and kurtosis of returns by assigning more weight to extreme negative returns. By incorporating downside risk measures in portfolio optimization, investors can construct portfolios that are more resilient to large losses during market downturns.

2. Consider asset correlation: The correlation between assets can be impacted by higher moments, such as skewness and kurtosis. For instance, a negative skew can lead to higher correlations during periods of high volatility, while a positive kurtosis can lead to lower correlations during these periods. Investors should consider how higher moments may impact asset correlations when constructing portfolios.

3. Use alternative weighting schemes: Traditional portfolio optimization methods, such as Mean-Variance Optimization, assume that returns are normally distributed. However, if the distribution of returns is non-normal, alternative weighting schemes, such as Risk Parity, Maximum Diversification, and Minimum-Variance Adaptive, may be more appropriate. These methods take into account higher moments and can improve the diversification of the portfolio.

4. Consider skewness and kurtosis when selecting assets: Investors may want to consider skewness and kurtosis when selecting assets for their portfolio. Assets with positive skewness may have higher returns during market upswings, while assets with negative skewness may provide some protection during market downturns. Similarly, assets with positive kurtosis may have higher returns during periods of volatility, while assets with negative kurtosis may have more stable returns.

Incorporating higher moments in portfolio optimization and risk management can provide investors with a more comprehensive view of the distribution of returns, leading to more robust portfolios and improved risk management. However, it is important for investors to carefully consider the impact of higher moments and to use methods that are appropriate for the specific investment strategy and goals.

## 6.10 Describe the concept of stochastic control and its application in algorithmic trading and risk management.

Stochastic control refers to the use of mathematical models to manage and optimize decision-making under uncertainty. In the context of algorithmic trading and risk management, it provides a framework for modeling and optimizing the behavior of financial systems that involve random variables and stochastic volatility.

In algorithmic trading, stochastic control helps traders make decisions based on historical data and current market conditions. For example, a trader may use a stochastic control model to determine the optimal time to buy or sell a security based on market movements and other relevant factors. By incorporating stochastic volatility into their models, traders can better anticipate market fluctuations and act accordingly.

Stochastic control also plays a vital role in risk management. For instance, it can be used to mitigate the risks associated with the holding of financial assets or securities. By modeling the uncertainty and volatility associated with such investments, traders can assess the probability of a potential loss or gain and make informed decisions about their positions. Additionally, stochastic control can enable traders to construct optimal hedging strategies to reduce the overall risk of their portfolios.

One common application of stochastic control in algorithmic trading and risk management is through the use of quantitative models such as the Black-Scholes model or the GARCH model. These models use stochastic calculus and other mathematical tools to predict future returns and volatility based on historical data and current market conditions.

To summarize, stochastic control provides a flexible and powerful tool for algorithmic trading and risk management by modeling and optimizing decision-making under uncertainty. By incorporating stochastic volatility into their models, traders and portfolio managers can make more informed decisions that mitigate risk and increase profitability.

# 6.11   What is the role of information theory in quantitative finance, and how is it applied to model selection and trading strategies?

Information theory plays a crucial role in quantitative finance and is widely applied in the development of trading strategies and model selection.

Information theory is a mathematical framework that deals with the quantification, storage, and communication of information. In quantitative finance, information theory is used to measure the amount of information contained in financial data and to extract valuable insights from it.

One of the ways information theory is used in quantitative finance is through entropy. Entropy is a measure of randomness or uncertainty in a system, and it can be used to quantify the degree of uncertainty in financial data. By measuring the entropy of financial data, researchers can gain a better understanding of the underlying patterns and behavior of financial markets.

Another important application of information theory in quantitative finance is to model selection. Information theory provides a useful framework for selecting the best model to fit financial data. This is done by calculating the information criteria of different models and selecting the one with the lowest criteria value. This ensures that the model is not overfitted to the data and can be used to generalize its predictions in unseen data.

Finally, information theory is also used in trading strategies. In quantitative finance, trading strategies are typically based on the analysis of data patterns, and information theory provides a useful framework for evaluating the strength of these patterns. By measuring the information content of financial data, traders can identify profitable trading opportunities and improve their decision-making process.

For example, suppose we want to develop a trading strategy based on the behavior of stock prices. We can use information theory to calculate the entropy of the stock price data and identify periods of high and low uncertainty. If we observe a high degree of uncertainty in the stock price, it may indicate that the market is about to experience a significant change, presenting a profitable trading opportunity. Similarly, if we observe low uncertainty, it may indicate that the market has stabilized, and we could consider implementing a longer-term trading strategy.

In conclusion, information theory plays a critical role in quantitative finance, providing a framework for model selection, trading strategy development, and risk management. Its applications are broad and varied, and it is an essential tool for any quantitative finance professional.

## 6.12    Explain the concept of dynamic copula models and their application in modeling multivariate dependence in financial markets.

Dynamic copula models are statistical models used to represent the dependence structure between multiple variables. They are particularly useful in financial markets, where the relationships between assets are often complex and ever-changing.

In a dynamic copula model, the marginal distributions of each variable are first estimated separately. This is because the behavior of each asset in the market may be different and can impact the overall dependence structure between the variables. Once the marginal distributions have been estimated, a copula function is used to model the joint distribution of the variables.

The difference between a static and a dynamic copula lies in how the dependence structure is modeled over time. Static copulas assume that the dependence between variables does not change over time, while dynamic copulas allow for time-varying dependence. This is crucial in financial markets, where relationships between assets are known to change over time in response to economic conditions, regulatory changes, and other factors.

Dynamic copula models help investors and traders better understand the multivariate dependence in financial markets, which can be used to manage risk and make more informed investment decisions. For example, by modeling the dependence structure between assets, investors can better assess the potential impact of a shock to the market, and adjust their portfolio accordingly.

One example of the application of dynamic copula models involves modeling the dependence structure between stocks in a portfolio. A trader might estimate the marginal distributions of each individual stock, and then use a dynamic copula model to model the joint distribution of the stocks over time. By doing this, the trader can assess how each stock is likely to behave in relation to the others, and adjust their positions accordingly to manage risk and maximize returns.

## 6.13   How do you apply the concept of market incompleteness in derivative pricing and risk management?

Market incompleteness refers to the situation where there are not enough tradable assets to replicate all possible payoffs. In other words, some payoffs cannot be created by combining existing assets (or their derivatives) in a portfolio. This poses challenges for derivative pricing and risk management because traditional methods based on no-arbitrage arguments may not be applicable.

One way to address market incompleteness is by using model-based pricing. This involves specifying a stochastic model for the underlying asset(s) and deriving the prices of derivatives as expected values under a risk-neutral measure. The risk-neutral measure allows for different probabilities of future events than the actual (physical) probabilities, in order to account for the possibility of arbitrage opportunities.

However, model-based pricing also requires making assumptions about the dynamics of the underlying asset(s) and the behavior of market participants, which can be challenging in the presence of market incompleteness. For example, the model may not capture all relevant risk factors or may underestimate the impact of large market moves.

Another approach is to use a combination of model-based pricing and hedging. This involves identifying a set of tradable assets that can be used to replicate some (but not all) of the payoffs of the derivative, and then adjusting the hedging strategy based on the model's predictions of market movements. This can help reduce the risk of the derivative position, but may also result in higher costs due to the need to trade in multiple markets.

A third approach is to incorporate some degree of subjective judgment or market views into the pricing and risk management process. This recognizes that market incompleteness may reflect differences in beliefs or preferences among investors, and that these differences can be exploited through active trading strategies. For example, an investor may use a combination of derivatives and underlying assets to express a particular view on the direction of a market, and then adjust the position as new information becomes available.

In summary, dealing with market incompleteness in derivative pricing and risk management requires a combination of quantitative modeling, hedging, and subjective judgment. The optimal approach will depend on the specific market conditions, the types of derivatives involved, and the investor's risk tolerance and investment objectives.

## 6.14   Describe the use of fractional Brownian motion and long memory processes in financial modeling.

Fractional Brownian motion (fBm) is a stochastic process that has been widely used in the modeling of financial market data. One of the main advantages of fBm is that it can capture long-range dependence or long memory in financial time series. This means that the autocorrelation between two observations decays slowly, implying that the effect of past prices on the current price persists for a long time. In contrast, in short memory processes, such as the standard Brownian motion, the autocorrelation between two observations decays rapidly, meaning that the effect of past prices on the current price is quickly forgotten.

Long memory processes are important in finance because they reflect the fact that market participants do not have perfect information and, therefore, they rely on past information to make their investment decisions. This results in a slow decay of the autocorrelation function, which corresponds to a slowly decaying probability of success or failure in trading strategies. The presence of long memory can also lead to volatility clustering, where periods of high or low volatility are followed by similar periods.

One example of the use of fBm in finance is in the modeling of stock prices. The Hurst exponent, which measures the degree of long memory in a time series, has been found to vary across different stocks, indicating that some stocks are more persistent in their behavior than others. By incorporating fBm into a model for the stock price, a trader can better estimate the value of the stock over longer periods of time, allowing for more informed investment decisions.

Another application of long memory processes in finance is in the modeling of interest rates. Long memory in interest rate movements can be captured by a fractional autoregressive integrated moving average (FARIMA) model, which takes into account the fractional differencing of the series. By incorporating long memory into interest rate models, financial institutions can make more accurate predictions of future interest rates, which can help them to optimize their portfolios and manage risk more effectively.

Overall, the use of fractional Brownian motion and long memory processes in financial modeling allows for a more accurate representation of financial market data and can help traders and investors to make better-informed investment decisions.

## 6.15  What is the role of behavioral finance in quantitative trading strategies, and how can it be incorporated into models?

Behavioral finance is the study of how cognitive and emotional biases affect financial decision-making, and it can be a very valuable tool for quantitative trading strategies. Many behavioral biases can cause inefficiencies or anomalies in the market, and integrating these biases into quantitative models may provide a significant edge in trading.

One of the most important biases in behavioral finance is the herd mentality, where investors follow the crowd rather than making rational decisions based on underlying fundamentals. This can create market anomalies that can be exploited by a quantitative trader. For example, a momentum strategy based on the fact that investors tend to chase past winners can produce positive returns over the long run.

Another important example of behavioral finance in quantitative trading is the concept of anchoring. Anchoring refers to the tendency for investors to place too much emphasis on a specific piece of information, such as a company's earnings forecast or the price of a stock relative to its historical average. This can create trading opportunities for quantitative models that take this bias into account.

Overconfidence is another bias that can be incorporated into quantitative trading models. Some traders believe they have an edge in the market and make decisions based on their own perceived abilities rather than actual market data. Models that incorporate measures of market volatility and risk can help to mitigate this bias.

Incorporating behavioral finance into quantitative trading strategies requires careful consideration and testing. The key is to identify the specific biases most relevant to the market being traded and to develop models that capture these biases in a meaningful way. Careful validation and testing are critical to ensuring the efficacy and robustness of any trading model.

## 6.16    Explain the concept of ambiguity aversion and its implications for financial decision-making and modeling.

Ambiguity aversion is a behavioral tendency in which people exhibit a lower willingness to take risks when the probability of an outcome is uncertain or ambiguous, even when the potential payoffs are high.

This is because individuals tend to have a greater preference for known risks, for which they can clearly understand and assess the probability of their consequences, compared to unknown risks. As a result, they perceive ambiguous situations as more risky than equivalent risky situations with known outcomes.

In terms of financial decision-making and modeling, ambiguity aversion has important implications. Firstly, it can lead to greater caution and conservatism in financial decision-making when there is ambiguity in the probability of different outcomes. For example, investment managers may prefer to invest in stocks or other assets with a known history or clear patterns, rather than in untested or uncertain products that offer potentially greater returns but also greater ambiguity.

Secondly, ambiguity aversion can affect how we try to quantify risk and uncertainty. Traditional quantitative models, such as the Capital Asset Pricing Model (CAPM) and Modern Portfolio Theory (MPT), often rely on assumptions of normal distributions, which may not always be applicable in the real world. As a result, these models fail to capture the full complexity of risk and uncertainty. Moreover, ambiguity aversion also calls into question the validity of using mathematical models in decision-making, given that these models can only use historical data as inputs and may not be able to account for future changes in risk factors.

Finally, ambiguity aversion can also have implications for risk management strategies, such as hedging and diversification. Ambiguity aversion can result in investors being overly cautious and missing out on potential opportunities, or being too conservative in their risk-taking, which can result in suboptimal outcomes. Therefore, risk management strategies need to take into account the impact of ambiguity aversion and be flexible enough to adjust as needed.

In summary, ambiguity aversion is an important behavioral bias in financial decision-making that can impact both individual choices and quantitative models. It is crucial for investors and analysts to be aware of this bias and consider it when making financial decisions, in order to achieve optimal outcomes.

# 6.17 How do you incorporate systemic risk factors in portfolio optimization and risk management?

Systemic risk is the risk that the entire financial system will fail or be disrupted, causing significant losses to investors and potentially leading to economic collapse. It's a crucial factor to consider when building a portfolio and managing risk, especially in today's interconnected global markets, where events in one corner of the world can quickly spread and impact the rest.

Here are some ways to incorporate systemic risk factors in portfolio optimization and risk management:

1. Diversify across sectors, asset classes, and geographies

Diversification is one of the most effective ways to mitigate the impact of systemic risk. By spreading your investments across different sectors, asset classes, and geographies, you can reduce the impact of any one event or region on your portfolio. For example, if you have a portfolio that is heavily focused on technology stocks, you could consider adding exposure to other sectors like healthcare or finance to reduce your overall exposure to any one sector.

2. Use stress-testing and scenario analysis

Stress-testing and scenario analysis are tools that allow investors to assess the potential impact of different events on their portfolio. By running various stress tests and scenarios, you can identify how your portfolio may perform in different market conditions and adjust your risk management strategies accordingly.

For example, you could simulate the impact of an economic recession on your portfolio and assess how much you may lose under different scenarios. This analysis could help you identify areas of your portfolio that may be particularly vulnerable to systemic risk, and adjust your allocation or hedging strategies accordingly.

3. Monitor macroeconomic indicators

Monitoring macroeconomic indicators, such as interest rates, inflation, and GDP growth, can provide valuable insights into the health of the overall financial system. By staying up-to-date on economic data, you can identify potential signs of systemic risk and adjust your portfolio accordingly.

For example, if inflation appears to be rising rapidly, this could be a sign of systemic risk as it could lead to higher interest rates and a slowdown in economic activity. In this case, you may want to reduce your exposure to assets that are sensitive to interest rate changes, such as bonds or real estate.

4. Consider alternative investments

Alternative investments, such as hedge funds, private equity, or commodities, can provide diversification benefits and potentially higher returns than traditional asset classes. These investments can also have lower correlation with the stock and bond markets, which can help reduce exposure to systemic risk.

However, alternative investments also come with their own set of risks and challenges, such as higher fees, less transparency, and liquidity constraints. It's essential to thoroughly research and understand these risks before incorporating alternative investments into your portfolio.

In conclusion, incorporating systemic risk factors in portfolio optimization and risk management requires a combination of diversification, stress-testing and scenario analysis, monitoring macroeconomic indicators, and considering alternative investments. By following these strategies, you can build a more resilient portfolio that can withstand systemic risks and potentially generate higher returns over the long term.

## 6.18 Describe the concept of optimal stopping and its applications in financial derivatives and trading strategies.

Optimal stopping is a statistical theory that deals with the decision of when to stop a process to achieve a maximum outcome. In finance, this theory is applied to decision-making processes where investors must choose when to stop a certain investment or trading strategy to maximize their returns or minimize their losses.

One classic application of the optimal stopping theory is in the pricing of financial derivatives, such as options. In options trading, investors have the right to buy or sell an underlying asset at a certain price within a specific time frame. The decision to exercise the option depends on the behavior of the underlying asset. The optimal stopping theory can help investors determine the optimal time to exercise their options.

In the case of a European call option, the optimal time to exercise the option would be the point in time when the underlying asset price is at its highest level. The reason for this is that the holder of the option will receive the maximum profit if they exercise the option at that point in time. Conversely, if an investor exercises the option earlier, they may miss out on potential profits if the price of the underlying asset continues to rise.

Another example of the application of the optimal stopping theory in finance is in high-frequency trading. In this context, traders must make split-second decisions to buy or sell securities based on market conditions. Using an optimal stopping rule, traders can determine the optimal time to make their trades and capitalize on market movements.

Overall, the optimal stopping theory is a powerful tool in financial decision-making, allowing investors and traders to make informed decisions on when to enter or exit positions to maximize profits or minimize losses.

# 6.19 Explain the use of agent-based models in simulating complex market dynamics and their applications in quantitative finance.

Agent-based models (ABMs) are computational models used to simulate the behavior of individual agents and their interactions within a given environment. In the context of finance, ABMs can be used to simulate complex market dynamics, incorporating different agents such as traders, investors, and institutions with different strategies, behaviors, and decision-making processes.

ABMs are useful in quantitative finance, as they provide a more realistic and detailed representation of the market, compared to traditional models that assume a perfectly efficient market. ABMs can capture important features of markets such as non-linearity, heterogeneity, feedback, and emergence, which are difficult to model using traditional methods.

One of the most notable applications of ABMs in finance is in the field of market microstructure, the study of the process of price formation and trading in financial markets. ABMs can be used to simulate the behavior of individual traders and their interactions in a market, providing insights into how market dynamics, such as liquidity and volatility, emerge from the actions of individual agents.

In trading and investment, ABMs can also be used to design and test trading strategies under various market scenarios. For example, traders can use ABMs to test how their trading strategies would perform under different market conditions, such as high volatility or low liquidity.

ABMs can also be used to study the impact of different policies and regulations on the market, such as circuit breakers and transaction taxes. In this way, quantitative analysts can use ABMs to develop insights into how the market would respond to different intervention measures, and inform policymakers on the potential outcomes of implementing different policies.

Overall, ABMs provide a powerful tool for simulating complex market dynamics and understanding the behavior of individual agents and how they interact in a market. By incorporating realistic features of markets, ABMs can provide valuable insights into trading and investment strategies, market microstructure, and policy analysis, helping to inform better decision-making in quantitative finance.

## 6.20 What are the challenges and limitations of applying machine learning and artificial intelligence techniques in quantitative finance, and how can they be mitigated?

There are several challenges and limitations when applying machine learning (ML) and artificial intelligence (AI) techniques in quantitative finance. Some of these challenges arise from the nature of financial data, while others stem from the complexity of the underlying financial system. Here are some of the major challenges and limitations of ML and AI in quantitative finance along with some ways to mitigate them:

1. Limited Data: One of the significant challenges in applying ML and AI techniques is the limited amount of high-quality data in the financial industry. Historical data is limited, and future market conditions cannot be known with certainty, particularly when there are structural changes in the financial markets.

Mitigation: One way to overcome this challenge is to use alternative data sources such as social media, satellite imagery, and other sources that can aid in predicting market conditions. Data augmentation techniques such as synthetic data generation can also be used to increase the amount of data available for training machine learning models.

2. Noisy Data: Financial data is often noisy and contains a lot of random noise, making it difficult to extract meaningful signals.

Mitigation: Techniques such as data cleaning, filtering, and feature selection can help to mitigate the impact of noisy data. Moreover, using models that are less sensitive to noise, such as ensemble learning models, can also be helpful.

3. Overfitting: Overfitting is a common challenge in ML because financial data tends to be non-stationary and has a high degree of randomness. When a model is overfitting, it learns the training data too well, becoming too specific to the data and losing its ability to generalize.

Mitigation: Some ways to mitigate overfitting include regularization, cross-validation, and using simpler models. Additionally, one can adopt a more cautious approach to model deployment by conducting stress-testing and backtesting of the model on unseen data.

4. Interpretability: Financial institutions are required to provide explanations for decisions based on quantitative models. However, most ML and AI models are inherently complex and hard to interpret.

Mitigation: One way to mitigate interpretability issues is to use models with explainable AI (XAI) capabilities. XAI models are designed to provide insights into how model decisions are made by providing reasons for the outputs. Moreover, one can also use sensitivity analysis and feature importance ranking to understand key drivers of the model's output.

5. Dynamic Market Conditions: Financial markets are dynamic and evolve continuously in response to various internal and external factors. Strategies that work well in one market may fail in another.

Mitigation: To mitigate this challenge, one needs to create models that are adaptive, self-learning, and capable of adapting to changing market conditions. One can use reinforcement learning techniques to create models that can learn from their environment and improve their action selection over time.

Conclusion:

ML and AI can offer significant advantages in quantitative finance, including improved prediction accuracy, risk management, and investment performance. However, several challenges and limitations must be addressed to successfully leverage these techniques in the financial industry. By improving the quality and quantity of data, using simpler and more robust models and ensuring interpretability, one can

more effectively mitigate these challenges and enhance the value of
ML and AI in quantitative finance.

Made in the USA
Monee, IL
08 January 2025

76352630R00075